LUST IN THE TRANSLATION

L~~U~~OST IN THE TRANSLATION

IT IS FUNNY NOW—IT WAS *NOT FUNNY* THEN

PABLO DIAZ

Printed in the United States of America and Canada

Interior Design: Creative Publishing Book Design

In Appreciation

TO MY HEAVENLY FATHER, who protected me when I was walking through the Valley of the Shadow of Death. Oh, how much I disrespected you, Lord. You overlooked my iniquities, ignorance, stupidity, and rebelliousness and still sent your beloved Son, Jesus Christ, the Ultimate Warrior, to come to fight for me, protect me, sometimes from my destructive self, and rescue me.

To those who took the ultimate step to leave our homeland, not sure of what to expect, and faced the trials and sacrifices that came with the steps we took in coming to the United States looking for better opportunities. Our struggles in learning a new language and many customs different from ours were well worth it. Our experiences were sweet and sour, but the sourest ones helped us along the way, and we, in turn, passed our "school of life" knowledge to those newly arrived in the country. I also want to express my sincere appreciation to those who spent a lot of time teaching me how to speak the language and understand a new culture. I express my deepest gratitude to those who mentored me and molded me into a better person.

To my three brothers Carlos, Felix, Luis, and two sisters Yarah, and Edilia, who also went through some of the same hardships as I did, as they moved to the United States in search of better opportunities. I would be remiss if I did not mention my father, Don Carlos, and my grandfather, Abuelo Antonio, whose stories drew a mental picture of something that I grew to like and then explore. And to my mother, Doña Aida, who took the time to show me how to do things and was always such a positive influence. You were "*loca, Mami,*" but I will always love you and miss you.

I would also like to extend my heartfelt appreciation and gratitude to Jose Manuel Marrero, my friend for life and brother in arms. Although we met after we had both retired from the Army, our experiences and trials as non-English speaking citizens of the United States coming out of Puerto Rico gave us an unbreakable bond. Your insight and mastery of written English helped me expand and better explain some of the stories with more detail and splendor, making them more understandable to those without any military background. Thank you, my brother, and may God continue to bless you and your family. I will be forever thankful.

Last but not least, to my best friend, mentor, and supporter—to my wife Kristi, who has been there for me and encouraged me to write this book. This is to the super mother and woman who has made uncountable sacrifices for those so dear to her. Thank you for being such a great mother to our boys Triston and Noah. They might not see it now, but as they grow older and live their own lives, they will see that your attention to their growing-up was not disciplining, but rather how much you care and love them. They will see how your encouraging advice provided the tools needed to mold them into great men; you are always doing what is best for them. With you, *"Best Wife Ever,"* I have finally learned the meaning of love.

Contents

Introduction

I AM NOT A FAMOUS person, and you have no reason for reading this book other than the somewhat catchy title, but I hope you enjoy reading it as much as I enjoyed writing it. For me, it was therapeutic!

Have you ever thought about your reason for being? Or the steps that took you to where you are and why things happened the way they did? Have you ever thought about your purpose in life? And what steps you needed to take to make you a better person?

Although this book is not intended as a recipe or formula to find a purpose in life or live a better life, many readers will find similarities between the personal experiences described here and their own. This book is about the peculiar things that happened to me during my introduction to the "Good Ole" USA as a member of the United States Army—things that are funny now but were not funny back then!

From my perspective, the things that happened to me were nothing less than terrible. Nevertheless, the incidents or situations that I experienced throughout the years molded me into the person I am today: an unconventional, think-outside-the-box, no-nonsense kind of guy. I have seen many things in my life that would turn a sane person into a blathering fool, some good situations, and some bad ones. Some of the bad ones were necessary for me to go through, as I found out later in life and saw what God had planned for me. Many times, I look back and laugh!

Now, this book is not just about my story or journey. It also puts in perspective and brings to life the uncountable tribulations of a non-English speaker going through the Army Basic Training (Boot Camp or Basic Combat Training), the Advanced Individual Training (AIT), and the unique struggles of "life in the Army." It illustrates the day-to-day communication challenges—many of which ***jeopardize*** people's lives. I believe that recruiters, drill sergeants, and all authorities associated with those "experiences" (training programs) will find plenty of "cautionary points" that could even save lives.

Throughout this book, you will also find the occasional use of what is sometimes considered derogatory language and offensive terms. The language and words are in no way addressed to anyone or displayed in a derogatory manner. Those segments in the book reflect the "Army English" that I was learning as I listened to the way people spoke around me.

They result from the natural process of learning a different language, as spoken in the Army, while making efforts to do my best to fit in the Army culture. It was my reality at that moment—take it as such and consider that many other individuals are experiencing similar trials and tribulations even as you read this work. And please keep in mind that I am not a Rhodes Scholar; if anything, I am more of a GED Scholar.

It Is a New World, at Least to Me

I LOVE MY HOME COUNTRY of Puerto Rico, but after I had lived there the first 16 years of my life, it was time for me to go and explore the world. There were no opportunities for progress on the island unless you were wealthy or well connected. Most of the time, the two terms worked hand in hand because you had the connections if you were rich. I had no ties to the wealthy crowd, so the opportunities were almost nonexistent for a guy like me who came from a low-income family. My only option at the time was to join the Armed Forces because my parents could not afford to send my siblings and me to college. This option was very appealing because my father, grandfather, and many of my uncles served in the U. S. Army, and I always liked to hear their stories.

I have always been fascinated with submarines and took the Armed Forces Test, intending to join the Navy and the Silent

Service. Because Puerto Rico is mainly Spanish-speaking, my dad had to teach me how to read and understand English. I did well during the test, as I could read English just as proficiently as Spanish. When I told the recruiter my desire to join the Submarine Force, he said that he would schedule me for an interview with the Navy Recruiter in San Juan, the capital, because the submariners were a very select, special group with particular requirements.

In the Spring of 1978, I reported to the naval office for the interview as scheduled, without knowing that the interview would be in English. As I previously stated, I could read but not speak English. The interviewer began to talk, and I looked at him like Bambi-looking at headlights. I did not speak one word of English. However, he kept talking as if I knew how to speak English. He then paused and asked me in Spanish if I understood what he was saying in English. I told him that I had no idea. He then gave me the shocking news that I could not join the submarine force; fluency in speaking English was a requirement. I tried to negotiate with him but to no avail. He told me that I could still join the Navy, but I would be on a frigate or even an aircraft carrier. I told him thanks but no thanks. If I could not join the Silent Service, that would be the end of my Navy experience.

Soon after that failed interview, I went to an Army Recruiting Station and joined the Puerto Rico Army National Guard. My dad had to sign papers for me to go on active duty, and he would not do so. My only option was the National Guard, and even then, he had to sign documents to approve my joining. He was adamant in his demand and told me so many times, "*if you don't like the military, at least you only have to deal with it one weekend a month.*" It did not matter to him that I was 100% sure that I would like it. I tried convincing him to no avail. Nevertheless, he had the power to approve/disapprove of my decisions. I was to leave Puerto Rico for the United States in a few months.

A few months later, in the summer of 1978, I left Puerto Rico, as scheduled, on my way to Fort Knox, a few miles outside of Louisville, Kentucky, for Basic Combat Training, or Basic Training as we used to call it. It was my first time off the island and my first airplane ride. I felt like Christopher Columbus as I was going to a New World—at least to me. And what a trip of a lifetime this was. The first leg of the journey was from San Juan, Puerto Rico, to Miami, Florida.

Due to this being my first time on a plane, I did not know what to expect. I was definitely scared, to say the least. Soon after we departed, the airplane began to make strange mechanical sounds, there was a smell of fuel mixed with oil, and weird noises echoed throughout the plane as it flew to Florida. The lights on the cabin were flickering, and the smell of oil and fuel grew more pungent, yet as scared as I was, I did not dare ask anyone what was going on so as not to sound as if I was stupid. It seemed normal to me since I had never

...the plane was having mechanical issues and would have probably crashed.

been on a plane before. I did not know how terrible things were until the flight landed in Miami. As soon as the wheels touched down, everyone got up and began to clap in celebration. The commotion was happening while the plane was taxiing down the runway extremely fast. I remember seeing people crying and hugging each other. I then asked the person next to me if it was normal for people to celebrate like this when a plane landed. The person told me that the plane was having mechanical issues and would have probably crashed, but the pilots did well dealing with whatever problems they were facing, and *gracias a Dios* (thank God), we made it!

Once outside the plane and in the Miami International Airport lobby, I noticed no language difference between the way people talked and behaved at the Puerto Rico airport and this location; it seemed like everyone was speaking Spanish. I said to myself; *This is not going to be that bad after all.* I asked others how to get to my next gate and waited for the next leg of my trip. I was soon boarding another airplane on my way to Atlanta, Georgia.

The flight to Atlanta was smooth and uneventful. It was nothing at all when compared to the one from Puerto Rico. Landing in Atlanta and walking throughout the terminal was more difficult and challenging than in Miami. It seemed like a bigger airport, and Spanish-speaking people were not as prevalent as in Miami. But I could still hear enough people speaking Spanish to comfort me that things would turn out okay. I had no idea how to get to the next terminal, but I somehow made it there after asking some Spanish-speaking people. Soon after that, I boarded the plane to Louisville, Kentucky. And this flight was also smooth and nothing like the flight out of Puerto Rico.

We soon landed at the Louisville, Kentucky airport. My first time at a location where I had no idea what to do and who to ask because I did not speak the language and Spanish-speaking people were nowhere to be found. Everyone who flew with me out of Puerto Rico went in a different direction. I thought to myself, *What have I done?* I began to look for any clues or a person who could tell me where to go or what to do next. As I started to panic, I came across two guys with what seemed to be Army haircuts and folders that resembled the one given to me by the recruiter in Puerto Rico. I followed them, hoping that they would lead me to my next location. They entered what looked like an Army bus, like the ones I saw in the movies, and I followed them there. The driver saw my folder and allowed me on the bus.

Soon the bus was rolling down the street; the drive went on for what seemed like hours. I started to panic because I had never seen so many "woods" in my life. You can go from town to town in Puerto Rico without noticing when you are in the next town. That was not the case here. The bus was going and going into the country without the sight of a town anywhere. I remember starting to sweat and question if I entered the wrong bus and worry about being lost in the United States, but not too long after that, I saw a sign that read, "Welcome to Fort Knox!" next to two Army tanks. Imagine my relief at that time. I later found out that the trip from the airport to Fort Knox took approximately 35 minutes. I was so nervous that it seemed a lot longer.

Welcome to the Army!

AS THE BUS ENTERED Fort Knox, I looked around for the building number in my Army orders—remember that I could read English but not speak it. The bus made a couple of stops where a few people got out. None of the buildings was the one listed on my orders. Suddenly, it stopped at this one building, and everyone that was still on the bus got off, except for me. It was not the building number listed in my military orders, so I remained seated. I remember the bus driver yelling in my direction. But I was sure that he was not talking to me, although I was the only one left on the bus because this was not my building number. He kept yelling in my direction, so I got up and showed him my orders with the building number where I was supposed to be. He pointed to the building next to the bus, and when I turned to look, he kicked me in the rear and off the bus. He then threw my bags right behind me. Yes, just like in the movies, this happened to me. I grabbed my bags and entered the building where a drill sergeant was yelling, and everyone was running in place or doing push-ups. *Welcome to the Army,* I said to myself. And I joined everyone else and started to run in place. What an introduction!

After a week at the reception station, squeezed with haircuts, vaccinations, issuance of uniforms, toiletries, and many other knickknacks, it was time for Basic Training. But not without a walk to another building. During this instance, we were digging holes for whatever reason. The drill sergeant gave me a paper and pointed to a building not far from where we were digging holes. I did not understand his instructions. However, I read the form, went in the direction that he pointed, and began looking for the building listed.

I entered the building and gave the paper to the person in charge. He gave me some documents and pointed to the desks. I then read the documents and noticed that it was a test. I waited for his verbal instructions, even though I did not understand them, and when he clicked on the timer, I began to take the test. I remember finishing the test, mostly English words and paragraphs. I gave the test with all the paperwork to the person and waited. He graded the test while I was there. I remember him writing and circling the number "92" in red. He then pointed to the door; I guessed he was sending me back to the barracks.

I went back to the barracks and remembered the drill sergeant asking, or I felt like he was asking me *something*, to which I nodded my head and shrugged my shoulders. He then sent me back to digging holes with the rest of the guys. Later, like years later, I found out that I probably took the English as a Second Language (ESL) Test. According to the test results, since I could read English, they probably assumed that I did not have to go to the ESL school, even though I did not speak English or understand anything that was said to me.

Basic Training photo. The photographer was probably telling me to smile. But me "No comprende."

This did not help me because, although I could read, the reality was that I did not speak the language. In hindsight, I realize that verbal communication was not only essential but critical during Basic Training; especially, while out in the range during "Camouflage, Cover, and Concealment" training. **Camouflage** is self-explanatory; **cover** is protection from hostile weapons fire; **concealment** is protection from observation. During this portion, we, the trainees, go from point to point, covering each other. One moves while the other provides cover by shooting at the enemy. If you are visible, of course, the enemy could see you and kill you.

So you can think of the importance of hiding from enemy detection and have cover from enemy fire. Suppose the terrain does not provide a safe, natural cover and concealment. In that case, you must maximize all the man-made materials available to camouflage yourself, your equipment, and your position.

What if I had stood up to run instead of crawl (because of the language handicap) and got shot?

In one area of the range, we had to navigate across a long ditch while someone with an M-60 Machine Gun was shooting at us. Imagine! We were to stay low and out of sight while crawling out of the area. I could feel dirt falling all over me as the machine gun bullets hit the edge of the ditch. Maybe there were some special effects, but it felt very much real. What if I had stood up to run instead of crawl (because of the language handicap) and got shot? Inadequacies in that respect could have cost lives, including mine - imagine that!

How does someone who does not speak English make it through Basic Training? By doing the monkey see, monkey do and thinking that "Querer es Poder," which translates to "where there's a will, there's a way." Because I did not speak English, something led me to be the last one in line at all the manual tests we had to take throughout Basic Training. By doing this, I had an opportunity to watch and practice the tasks before it was my turn to perform them. Because of this, I received a first-time "go" in all of them. Due to my inability to speak English but my ability to read and understand what I was reading, I spent my free time reading the regulations and manuals that we needed to read and would be tested. This made it easier for me when it was time to take a written test in any of them, and as a result, I did very well in all the tests.

Many of the teachings in Basic Training were extremely challenging because of the language barrier. I immediately forgot many of the subjects. Thank goodness I re-learned all of them during future assignments as I learned to speak the language. These are just a few of the numerous funny events that are funny now but were *not* funny then.

I did not understand it at the time, but whenever the drill sergeant yelled for me or at me, he almost always said *mother f***er Puerto Rican*. So why did I take it? Because I did not know what mother f***er meant at the time. I understood who he was talking to when he yelled Puerto Rican because it sounded like Puerto Rico. I was the only one in the group from Puerto Rico. In fact, I was the only one that was not black or white in the entire platoon. So I was sure he was talking to me.

One day I was sitting on the ground, cleaning my weapon upon returning from the shooting range. The drill sergeant was yelling at another Soldier and yelled, "MOTHER F****er." Upon hearing it, I quickly jumped to my feet, snapped to the position of attention, and my rifle, along with the parts that I was cleaning, flew all over the place. I answered, "Jes, Drill Sergeant." I stood there for a few seconds and slowly turned in his direction. The drill sergeant and the other guy were looking at me like I was crazy. I slowly said, with my strong accent, "Jew call me drill sergeant?" He looked at me and, after a few seconds, continued to yell at the other guy. When he did not come my way, I collected the rifle parts, sat back down, and continued cleaning my weapon. I remember sitting on the ground and thinking that the drill sergeant better not come over to me since I got up when he called me. As you already know, it is funny now, but it was not back then!

Here is another one of the incidents. I got into a fight with a guy who went by the last name of Peterson. For some reason, he kept picking on me. He would either yell something to me or push me. I would be walking in the barracks, and he would push me. If I were trying to go somewhere and he was near, he would block my way. He pushed me a few times for no reason. If we had to do something together, he would make it difficult or do something to make me look like a fool. No matter what we were doing, he would do or say something every time he was near me, and everyone near us would laugh. I do not know why; he just did something. I grew tired of the way he was treating me.

One day, while I was practicing commands with the guidon (the guidon is a company flag or color), Peterson pushed me and took the guidon from me. This was the last straw, so I pushed him back and re-took the guidon, and when he tried to take it back, we got into it. The rest of the platoon came to separate us. He was yelling at me in English, and I was letting him have it in Spanish. I believe that I got my point across, and he understood that I would not be putting up with his bullying anymore. Later that day, another platoon member called me to the side and told me, "Next time, you tell him suck my Peterson." I did not know what it meant, but I practiced saying it. When the opportunity arose, I said it. I looked at him straight in the eyes and said, "Suck my Peterson," even though I did not know what it meant, other than his name was in the phrase. I did it to the delight of the rest of the guys in the platoon. They were laughing so hard, and he looked so embarrassed that I felt terrible for him. That was the last time I had to deal with Peterson. Sometime later, as I learned how to speak English, I understood what the comment insinuated and thought it was catchy and funny. Poor Peterson!

Basic Training was full of "Army experiences." Lights out was one of those experiences. I remember that lights out was at 10:00 PM (or 2200 hours in Army language). That meant to be quiet and go to sleep. This one I learned from my dad, who did a tour in the Army during the Korean Conflict. The only two authorized to be awake after 10:00 PM were those on Fire Guard duty. Since the barracks were made from wood and incredibly old, two people always needed to be up while the rest were sleeping so that they could wake everyone up in case of a fire. The whole building could burn to the ground in a matter of minutes, something else that I learned later but did not know at the time—yes, when it was not funny.

Every day and every night were full of surprises. On one occasion, after lights out, a few of my fellow trainees kept talking rather loudly and making a lot of noise. They were joking, and I could hear their laughter. Unbeknownst to us, the drill sergeants were outside, also listening, and out of the blue, we were abruptly awoken at 2:00 AM and rushed outside in our underwear to do push-ups on the gravel. I remember the drill sergeants yelling at every single one of us. Who knows what they were saying, but they laughed, and we did not think it was too funny. Once THEY got tired, we were rushed back to our bunks, dragging dirt all over the floor in the barracks and our bunks. Something else to add to our already busy schedule before barracks inspection in the morning: Clean all the dirt we just dragged in. You know, since we had a lot of free time, which we did not. I can imagine that they probably told us to keep our mouths shut and go to sleep like we were supposed to do. From then on, you could hear a pin drop after lights out.

One of the most significant differences is that as a civilian, you can question orders and authority; there is no room for that in the military.

Let me pause to make a point crystal clear. Basic Training was a grueling, life-changing transformation. It was hell—even for those who spoke perfect English. Words like trepidation, anxiety, frustration, and pain fall short in describing what a trainee experienced back then.

Basic Training is where the military psychologically and emotionally transforms you into a warfighting machine. The military pushes the "reset button" in your mind as soon as you report to Basic Training and begins to change how you are supposed to act, work, and think.

One of the most significant differences is that as a civilian, you can question orders and authority; there is no room for that in the military. It is an entirely different world.

The critical elements instilled in you, directly or indirectly, are assertiveness and courage. It is worth noting that the military does not teach *Humble-Confidence.* Let's face it. No effective army will stand an enemy with a pusillanimous attitude. Unimaginable! In that view, you are toughened from the first moment in Basic Training. Trainees are instilled offensive enthusiasm, displaying self-confidence. Whether you like it or not, you are expected to have the highest standards in all you do—even in the minor tasks. You are also taught to overcome your fears and any natural clumsiness. The result? An individual who displays self-confidence and determination to handle any job, anywhere, and anytime.

I have spoken with many civilians who always ask me why Soldiers are so arrogant and possess such an aura of invincibility. I respond that is the way to survive when you are in the military—not to mention what I have explained in the previous paragraphs.

Imagine not having the ability to talk with anyone because of the said language barrier—no one to voice your fears, worries, and concerns.

There is no room for weakness in the military, as it would cost your fellow Soldiers' or your own life. As a civilian, when you receive an order, you can question it and walk away if you do not want to do something. In the military, as I stated, that is not an option. Challenging an order could result in death. There is no switch to turn it off and on as we go home for the day and interact with civilians during our free time away from the military. You are trained to anticipate the worst-case scenario: death. So you plan, coordinate, and execute accordingly. Civilians might call this paranoia; we in the military call it awareness. It is a way of life. You leave Basic Training with a "new mentality," and, in fact, it gets nurtured as your career progresses.

Imagine, again, what Basic Training would be like for someone who is learning Army skills and English at the same time. Just think of what goes through that trainee's head when he has no clue what is being said around him/her. Words and commands are barked left and right, and you do not understand anything at all. You are in total isolation. It is a new world.

Imagine not having the ability to talk with anyone because of the said language barrier—no one to voice your fears, worries, and concerns. You suffer the uncertainty of what was going to happen next. You constantly worry about whether you will have the ability to complete any given task. You do not know how to ask someone to explain what was just taught to your platoon.

All trainees suffer stress—no doubt. However, that stress is duplicated by the inability to communicate with others. It is a miracle that I was able to make it.

In one instance, I was told or yelled at to report to KP. I had no idea what KP meant (Later, I found out that KP stood for kitchen police.) At that moment, I was wondering—Was it guarding the kitchen? It seemed to me that there was somebody there all the time. What would I be defending or protecting? So, off I went with another guy from my platoon.

We reported, and, once again, I did not understand what was going on. I followed the hand directions of the cook in charge. He gave me a potato peeler, led me to a couple of barrels filled with potatoes and water, and showed me how to peel potatoes. I have never before or since peeled that many potatoes in my life. But because "I no comprende" (which is not proper Spanish), it was my luck to do things of that nature—the crappy tasks that nobody wanted to do. Yep....give them to the non-English speaking dummy who has no idea what is going on.

However, I did enjoy cleaning the pots and pans; it was my favorite chore growing up in Puerto Rico, not to this level, but I liked to wash the dishes back home. This was when I looked around to make sure nobody saw me and ate some of the food I was supposed to be throwing away before I cleaned the pots. I think the cook saw me eating a couple of times and yelled at me, but he did not stop me from eating. The next time I heard the Drill Sergeant yell KP, I raised my hand. It was great—peeling potatoes and sneaking and eating leftover food from the earlier meal! Well, needless to say, I ended up loving KP.

One thing that I loved while in KP was the music. The mess sergeant kept a radio on while they were cooking and we were cleaning. I remember hearing many songs (Although I did not understand them) that were captivating. Years later, I figured out that I was listening to:

"Blinded By The Light"
Manfred Mann's Earth Band

"Two out of three ain't bad"
Meatloaf

"Reminiscing"
Little River Band

"You Make Me Feel Like Dancing"
Leo Sawyer

"Three Times a Lady"
The Commodores

"How Deep Is Your Love"

Bee Gees. (More about the Bee Gees in another segment.)

"Baker Street"

Gerry Rafferty

"When I need You"

Leo Sawyer

"Too Much Too Little Too Late"

Johnny Mathis and Deniece Williams

"Misty Blue"

Dorothy Moore

If you have heard those songs, I am sure you love them as much as I do. The curious thing, I repeat, was that I loved them but did not understand the lyrics. I just could understand ***some*** of the words because, for the most part, they were not fast-paced songs.

There was something special about those songs. Perhaps I have an ear for music. I come from a musically inclined family. My parents had a one-hour radio show when they lived in Tampa, FL, and traveled all over Florida singing old Spanish songs. Many of my uncles played with bands in Puerto Rico. I tell you, people who were blocks away could hear me beating the Latin Percussion (LP) congas.

Okay, back to the harsh and unforgiving realities of Basic Training. Let me go over some of the daily "activities" during that time. I did so many push-ups in Basic Training simply because I could not say 'Yes.' It just would not come out. What came out was 'Jes,' no matter what: 'Jes, Jes, Jes.' Here is the thing: Before every meal (breakfast, lunch, and dinner), we had to do 20 push-ups, the monkey bars, then stand in front of the drill sergeant and answer to whatever he said with "Yes, Drill Sergeant." It did not work out like that for me. It was more like "Jes, Drill Sergeant." And as soon as I said 'jes,' he showed his hand with the thumbs down, which meant 20 push-ups, and sent me to the end of the line to repeat the drill. Here I go, 20 push-ups, monkey bars, and 'Jes, Drill Sergeant' one more time. This happened a few times before he felt sorry for me or got tired of sending me back. There was no way I could say 'Yes'—it was 'jes.' I would lie in bed at night and try saying it, but it would not come out, and I remember getting frustrated and punching the mattress in the bunk bed. I am sure the guy on the bunk below was not too happy; he kicked the bunk bed a few times and said something, probably telling me to stop. It all came to an end when I realized that if I said the letters E-S extremely fast, it sounded like 'Yes.' I could not wait until the next morning to put it into practice.

The next morning came, and when I did my push-ups and monkey bar, I answered, "ES Drill Sergeant." The Drill Sergeant looked at me funny and waved me through. I never learned what he was saying, but I knew that when I answered with a resounding "ES Drill Sergeant," I could go inside the mess hall and eat with the rest of my platoon!

After weeks, going through the food line and eating while getting served (because I was running out of time and could not sit with my group), I could finally sit down with the rest of the guys in my platoon and enjoy my food. Maybe this is where I picked up the bad habit of eating fast, which I have up to this day. I have tried to slow down many times, but always go back to eating fast. Thank you, 'JES, DRILL SERGEANT!'

Another experience from Basic Training was the famous *blanket party*. Something I learned after the fact as I reflected on Basic Training later in my career. A blanket party is a form of corporal punishment, hazing, or retaliation conducted by a peer group, most frequently within the military. The victim, who is usually asleep in bed, is restrained by having a blanket thrown over him and held down, while other members of the group strike him with a sock or bath towel containing something solid, such as a bar of soap. One of our fellow trainees was not doing what he was supposed to, such as shinning boots or making the bed up to standards, and because of this, the rest of us found ourselves cleaning up after him. Thus, we were chastised for his mistakes and lack of focus. I suppose that was the drill sergeant's way of sending a stern message across regarding "combat cohesiveness." During combat, when one person makes a mistake, other Soldiers, not just that individual, get killed. Since we were molded as a combat team, we had to understand and *feel* that collective responsibility or 'collective combat consciousness,' as a friend calls it.

On one occasion, after returning from a long and gruesome forced march, on a rainy day, a miracle happened. The drill sergeant felt "compassion" and gave us the rest of the day to get our stuff together (clean our muddy boots and have our area ready for an inspection a few hours later). I do not know what this guy did the entire time, but I remember the drill sergeant coming to inspect. The first bunk he inspected belonged to—yes, you guessed it—this guy. It was a real mess.

The drill sergeant got extremely upset. He found his boots on display but soaked and full of mud. So the drill sergeant kicked a smoke can all over our shiny waxed floor. Subsequently, he yelled for two guys to go with him. I can only assume that he told them about doing a blanket party on this guy. Upon cleaning the mess created by our careless fellow trainee, it was soon time for lights out.

A few hours later, the same two guys that the drill sergeant pulled aside woke us up under cover of darkness. They led us to the guy's bunk and whispered to keep quiet. I did as told and saw a blanket over the guy's bed with him trying to get off from under it. The rest of the guys were punching him and hitting him with what seemed like socks. There was no way I was going to punch him, but I also thought that I might be the next victim if I did not. So instead, I chose to hit the mattress a few times and return to my bunk. As you can imagine, the poor guy was wailing and groaning the rest of the night. The next morning, we could see his bruises and a black eye. We did not have any more issues with him during the rest of Basic Training.

So what did I think of drill sergeants after all? Tough and experienced leaders who were preparing us, those with no experience, for a tough life in the US Army! They had to be that way!

The drill sergeants were so sharp and knowledgeable that I wanted to be like them. I thought that if I had an opportunity, I would try to be one later on. This group is responsible for coaching and mentoring thousands of Soldiers and transforming them from civilians to warfighting machines.

Drill sergeants are a class apart. They are distinguished warriors and instructors who earned that title. They can also be identified from far away by their unique hats and how they speak with such authority, command presence, and pride. Many of us dreamed about becoming one of them, but I really felt that I was born to earn that title and unique responsibility later in my career.

What happened to me after Basic Training? Soon after graduating, I went to Fort Harrison, Indiana, for Advanced Individual Training (AIT). Fort Harrison was a completely different experience. During this time, mostly during my first month there, I ate nothing but hamburgers and hot dogs for lunch and dinner. Not because I wanted to, but because I did not know any better and did not ask anyone; I still did not speak good enough English. I saw people with my same rank going to the line that only served hot dogs and burgers. The ones going to the other line were mostly sergeants and officers. Eventually, I saw a Spanish-speaking guy from my unit during lunchtime, and he asked me to go with him to the other chow line. I told him that I could not because I did not have enough rank to go to that line. He told me that it was for everybody. Imagine my surprise when I saw pork chops, chicken, stew, and a lot of other tasty food.

In retrospect, I realize the stupid things that I did because I did not speak the language or bothered to ask others who spoke my native language.

I remember this one time, a female trainee asking if she could sit with me or if I minded if she sat with me during lunch. I meant to say yes but said, “Yes, I mind.” She got upset and left to sit somewhere else, giving me a dirty look. I did not have the words in my vocabulary to tell her that she could sit down. I met a lot of friendly people at Fort Harrison and wished that I had stayed in contact with them. Some of them took the time to try to teach me how to speak English. We spent countless hours speaking “Spanglish.”

Even though I did not speak English, I learned many Army cadence songs during Basic Training. I remember a particular song that I sang during my first opportunity to call cadences during physical training at Fort Harrison.

Keep in mind that I was repeating whatever I heard drill sergeants and others singing. The Fort Harrison group, unlike all males at Fort Knox, consisted of over half females. I did not know the meaning of the words in the cadences and just repeated what I heard. Cadences are songs that the one person in charge of the group sings and the other 40 or so Soldiers in the formation repeat. Thus, the words go exceptionally far when repeated by the group.

This was one of the most familiar cadences. There are many variations to it, but the one I sang went like this:

C130 rolling down the strip;
Airborne daddy gonna take a little trip.
Stand-up, hook-up, shuffle to the door;
Jump right out and count to four.
If my chute don't open wide,
I got a reserve by my side.
If that one should fail me too,
Look out world I am coming through.
If I die on the old drop zone,
Bury me in a Russian *cunt.*

Senior Instructor: *Private, are you crazy? You cannot sing that, not in front of all these women.*

Private Me: *Huh? What I say?*

I am sure there were many people at Fort Harrison who should not have heard that cadence (not the way that I sang it). To this day, they might still be scratching their head, saying, "What did they say?" That was one of the many times they had somebody translate and explain what I sang. It is funny today, but it was not funny then!

Another unforgettable experience happened on a Saturday night when I got sick. My stomach was hurting badly. Thus, some classmates took me to the Fort Harrison hospital emergency room. I had an excruciating pain that was killing me. It was not food poisoning. The authorities did some tests but could not find anything wrong. So they decided to do a rectal examination. I had no idea what it was at the time. Still barely speaking English, I had no idea what was about to happen. The doctor said something to me, but of course, I did not understand, so he grabbed me by the waist and started pulling down my pants. I pushed him away, and the male nurses tried to grab me, and I also pushed them away. They tried again, and I started to kick and swing at them. I guess they tried to explain what they were trying to do but, once again, I was not having any of it. They then grabbed me, flipped me over on the bed, tied me with some belts, pulled my pants down, and did what they needed to do. I was cursing at them in Spanish the entire time. Even though I was lying on my stomach, I tried to punch them. It was not a happy time. They finally gave me some Maalox and sent me back to the barracks. I remember still cussing at them on the way out of the emergency room.

As if that traumatizing experience was not enough, they also gave me some pain pills that made me drowsy. So when I got to my room in the barracks, I took the pills and went to bed. One of my roommates thought I was sleeping hard due to my illness, so he brought a girl to the barracks room and had sex with her while I was asleep—or so they thought. They did not respect each other or were so horny they did not care if there was anyone else in the room. We had bunk beds, and I was sleeping on the top bunk. I did not see anything, but I also did not want to hear anything.

Fort Harrison was a nice base, and Indianapolis is beautiful. Still, after eight weeks, it was time to go back to Puerto Rico to begin my duty as a Finance Specialist with the Puerto Rico Army National Guard.

Back Home in Puerto Rico

HERE I WAS, BACK where I started, and wishing that I were somewhere in the United States with all the people I met during my training, but that was not possible. It was time to start my career with the Puerto Rico Army National Guard. I missed learning to speak English. Although it was hard, I missed the challenge of learning the language.

During this time, the movie Saturday Night Fever was still hot. I took a liking to the songs, and the Bee Gees became my favorite group. I remember getting an eight-track tape with the movie's soundtrack and singing the songs with my friend Gerry Vaello. I tried extremely hard to assimilate their voices, and because I liked them so much, I practiced English by memorizing and singing their songs. However, here was the funny thing. The Bee Gees members sang with a high-pitched voice known as falsetto. Thus, when I was singing "along" with them, I tried to make high-pitch sounds, which was quite unnatural for me.

Another funny anecdote. San Juan is the capital of Puerto Rico and is across the bay from my hometown of Cataño. There is a single route ferry that we call La Lancha de Cataño (Cataño Ferry), and it transports people back and forth between both towns. As such many times, tourists wandered to our neighborhood looking for the Bacardi rum plant. The plant was within walking distance of us. During these very few interactions with tourists, I gave them directions to the plant in my limited English but in a high-pitched voice. Yes, like in the Bee Gees songs. Picture this, I am telling a tourist in my best high pitched voice to "*take this road, make a left at the corner...*" Can you imagine that? Thinking about the Bee Gees, you know you want to do it, come on, sing it, "Ah, ha, ha, ha, stayin' alive, stayin' alive. Ah, ha, ha, ha, stayin' alive, stayin' alive."

My time with the Puerto Rico Army National Guard was very disappointing. There was no discipline or *esprit de corps* there. Every time we went on a weekend drill at Camp Santiago, Salinas, Puerto Rico, it was more like a camping trip, and the order of business was drinking rum and playing cards. However, I must add that when we went on the two-week Annual Training, this was the most formidable working bunch of people I have ever met. I do not know when the transformation took effect, but their performance was unparalleled during this time; the regular once-a-month drill routine was not appealing and very disappointing.

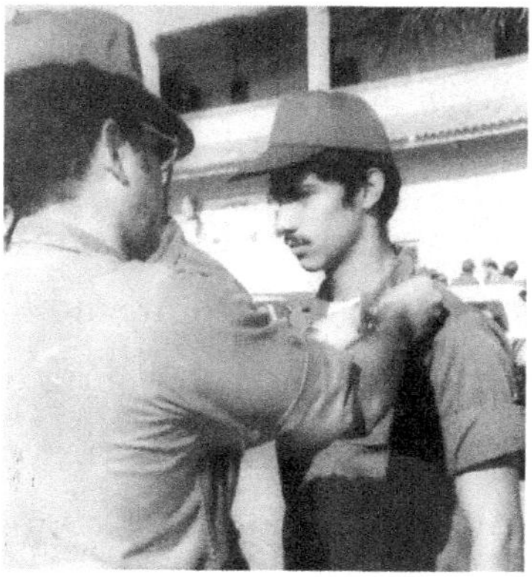

After spending well over a year in the unit I finally got promoted from E-1 to E-4 in one weekend. However, that was too little too late.

When we were not at Camp Santiago, we were at the Headquarters based in San Juan and did nothing but sit around to kill time. Only those that were part of a select group got the best opportunities and, as you can imagine, moved up in rank faster than the rest of the unit.

Although I did everything that was asked of me and never missed any of the drills or training, well over a year after returning from AIT, I was still a Private-One; no promotion at all. It seems like I was the forgotten man. The weekend I turned in my request to go on Active Duty, the Colonel that had to approve my paperwork decided to promote me from E-1 to E-4. Yes, a 'multiple-level promotion." He apologized and had a long conversation with my chain of command.

I accepted the promotions; there was no other option. Besides, I was used to following the orders of those appointed over me; in this case, of course, it was easy. At the same time, I felt different about this one. Honestly, I couldn't care less.

The promotions did nothing for me because once I joined the Active Army, two weeks later, I was dropped back to the rank of PVT, E-1, per regulations. So, I had to start from scratch all over again. Thank you, Colonel, for admitting that my entire chain of command forgot about me. Also, thank you for your efforts to right a wrong. You tried to do what was ethically and morally correct. However, it was too little, too late. In any case, off I was on my next adventure!

Back to the Army

TWO YEARS AFTER RETURNING to Puerto Rico from Fort Harrison, Indiana, I finally did what I wanted to do from the beginning: join the Active Army. On January 22, 1980, I left the island on my way to Fort Jackson, South Carolina. I vividly remember the big difference between Fort Harrison and Fort Jackson. Fort Harrison was like going to college; Fort Jackson, like going to jail. There was no freedom, and it was worse than Basic Training at Fort Knox, Kentucky.

Type, type, type, tic, tic, tic. Dang, my fingers hurt. I had never laid my hands on a typewriter, but this was typing school, and I had to pass this course in a month with the goal of typing 20 words per minute (wpm), or my future job in the Army was going to get rearranged quickly to fit the famous "*needs of the Army.*" I had always been a good student, so I put my best foot forward. I typed well over 25 wpm in about three weeks, even though I had no idea what some of the words or sentences meant. I memorized, "now is the time for all good men to come to the aid of their country." But I questioned, "Isn't that why we joined the Army?" They kept trying to explain that it was a sentence that used most of the keys on a typewriter and, thus, a particularly useful typing drill. Okay, whatever...that was me and my inquiring mind getting in the way of learning.

When it was time for the final test, no matter how hard I tried, I could not type 20; it was 18, 19 but not twenty. The instructor understood that I was probably nervous, so he took one of my earlier practice tests and used it as my final.

I remember that the drill sergeants would often have us outside the barracks in formation at 5:00 AM, in the snow sometimes, while they were inside drinking coffee and joking for over an hour before marching us to breakfast; we could hear, and sometimes see them from the outside. I never understood their rationale, but it was the old Army, where some things did not make sense.

Our living accommodations were not luxurious. And unlike what a fellow Army Recruiter told an applicant later in my career, this was not like college dorms at all. There were eight of us men sharing a room in the barracks. In my case, it was seven African-Americans, and yours truly. I still did not speak English well but paid a lot of attention to whatever they were saying to learn the language. I remember them using the word 'n' back and forth when speaking with each other. The term "n" is understood as a derogatory word used by some to address the African-American population. Due to its sensitive nature, still to this day, I decided not to use the word and instead use "n" in the story.

Among my African-American roommates, the "n" word was used extensively daily. It was 'n' this, 'n' that. I heard it so much that I thought it meant buddy or friend. Remember, I was learning a new language, which was the first time I heard that term. I was also learning about the many cultures in the United States. Additionally, I reiterate, it was a word that my roommates used quite a lot, without feeling insulted. Evidently, I assumed it meant "buddy" or something like that.

A few weeks after starting our training, we all received a day pass (about four hours) on a Saturday. My roommates decided to stay in to play cards with two other guys from down the hall. In my effort to be friendly, I walked to the Shoppette, a convenience store inside Army and Air Force bases. I bought some chips and sodas and walked back to the barracks. I practiced what I was going to say to them as I was walking to the room. I wanted to make sure I was cool and show my roommates that I was paying attention and learning the language. I mean, I really did practice this aloud as I was walking back. Once I arrived at the barracks, I was ready to impress them! I opened the door to our room, lifted the bags in the air, and said, "What's up, n******?" One of my roommates fell to the floor from laughing so hard, but the two guys from down the hall did not think it was funny at all and tried to kick my butt.

Thankfully, my roommates got between us and explained to them that I did not speak English well. Oblivious to what was going on, I kept using the word and questioning them, with my limited English vocabulary, about what was going on. I remember telling them that they called each other "n" so why was it wrong for me to say it? They never told me what it meant and only told me not to repeat it to anyone. I did not sleep at all the rest of that weekend, waiting for those two guys to come back and beat me; I was thinking blanket party on me. Every time I heard some noise in the middle of the night, I woke up scared.

All of this because I called them what they were calling each other, a word they used all the time and taken in a positive.

All of this because I called them what they were calling each other, a word they used all the time and taken in a positive. So, again, I thought it meant buddy. They still would not tell me what it meant and only told me not to repeat it. Believe me, I did not use it anymore. I learned and used the other words: buddy and yo!

Let me highlight the relevance of learning both language and culture. When you learn another language, you must learn about the culture as well. Music is part of it. For instance, my roommates were always singing about Mrs. Jones. And I often wondered what the thing was they had going on! It was all the time, "Mrs. Jones, Mrs. Jones, Mrs. Jones." She must have been incredibly special—I thought.

I was also very naïve at the time. Eventually, I learned it was a song, "Me and Mrs. Jones" by Billy Paul.

Learning English was one of the hardest things I had to do, and to this day (over 30 years later), it is still hard. One of my first experiences trying to speak the language was when two of my classmates (a male and a female), who took an interest in helping me, wanted to go to dinner on a Saturday night as a group. After six weeks, we were finally allowed to wear civilian clothes and leave the base. I hesitantly agreed because I was not well versed in the language, and it was still difficult to communicate with others.

I was watching TV in the day room while waiting for them. During this time, I saw this commercial where a man told the woman, "You look terrific, and she smiled." This was the first time any of us had worn civilian clothes, so when I saw my female friend in civilian clothes trying to be friendly, I repeated what the guy in the commercial told the woman. Remember, in the commercial, the guy told the woman she looked "terrific." However, I said, "You look terrible!" I am sure I was smiling when I told her, but that was not what I was trying to say. It took me a long time to convince her that I meant to say, "terrific." The things that we learn and the way we are exposed to them.

On another occasion, I was sitting next to a trainee on the 24-hour duty called Charge of Quarters (CQ). I was using the desk as a drum. I pretended to be playing *congas* (modified drums for Latin music) on the desk and singing Spanish songs. The person was telling me to go away, but due to my limited English vocabulary, I did not understand, so I kept at it. A drill sergeant happened to walk by, and the CQ told him what was going on. The drill sergeant nicely told me, "Private, go on your way." I looked at him, puzzled because I did not understand what he was telling me. He repeated himself, trying to be nice once again. I still did not know what he was telling me. So he raised his voice and yelled, "GET THE HELL OUT OF HERE." I replied, "Yes, drill sergeant!" with a smile on my face. I was not offended; this was the language that I was used to hearing all the time. I left the area, and that was the end of that.

Imagine attending a class when you do not understand most of what is said and taught, for that matter. Well… that was my situation. I had to rely on reading. It was more difficult when I did not understand something and had no way to ask questions. I remember feeling anxious during tests. Fortunately, I felt relieved after the results came back. I passed the course! To this day, I still do not know how I did it. It must have been our Father in Heaven who—although I did not want to acknowledge, was helping me—He was, is, and will be omniscient, omnipotent, omnipresent, and omnibenevolent. He was there for me every step of the way. But I ignored him. Why was I so ungrateful?

Soon after, we graduated. I was on my way to my next destination: Fort Stewart, Georgia.

Fort Stewart, Georgia

I HAD NO IDEA WHERE Fort Stewart was and how I was going to get there. I was expecting another plane ride. But it was not so. We all marched to the travel office to pick up our tickets. Everyone, as is the military way, was going to different duty stations. Some were going back home to their National Guard and Army Reserves units, which was my case during my Ft. Knox and Ft. Harrison days. But most of us were going to an Army base somewhere in the United States or overseas, such as Germany or South Korea. In my group was a Soldier who had been in the Army for a few years who was changing military classification (job). The drill sergeants put me under his leadership for traveling purposes and told him to make sure I was where I needed to be. We were the only two from the entire class going to Fort Stewart, Georgia.

Not too long after that, we packed our stuff, cleaned the barracks, and went to bed to wake up early the next day and go in our separate directions. After the graduation ceremony, all Soldiers left for their "permanent party" (actual assignment or job, not training) locations. I did not even know why it was called permanent party. At any rate, I was following orders; we got on a bus that left Fort Jackson at around 10:00 AM.

It felt like we stopped every 30 to 40 minutes to pick up or drop off someone. A few hours later, we stopped at a bus station for a couple of hours. The Soldier in charge of me waved, indicating to follow him. He took me to a diner, which was great because I was starving. Everything seemed to be going well. I tried to speak English with him, but it was still difficult. This time at least, I had a few words that I could try to put together. Soon enough, a waitress came to take our orders. He ordered first since he knew how to speak English and what he wanted. When it was my turn, I told the waitress, "I want a God-d***ed hamburger (Forgive me, Lord, You know I did not know what I was saying) and some mother f***ing fries." I said this enormously proudly, as it was probably my first time placing an order in English. They both looked at me like I was crazy. The waitress gave me a dirty look, and my fellow Soldier asked me why I was talking like that. I did not understand what he meant and told him, "I speaking English," not understanding that the type of language I was using was vulgar.

To this day, I still do not know why foul language is used in the military, but it was the norm. Why? Because it was the first time, many of us were away from home and want to show that we were adults? Mom and Dad are not around, and now we can say whatever comes out of our mouths without having to worry about getting in trouble or grounded. Why? Because we thought it could help to endure physical and mental pain? I do not know. But the language was pretty much standard and prevalent. Nevertheless, that was the language I learned and used because it was what I heard all the time. Later, when we stopped for dinner, he ordered for me!

Here's a picture of me later in my career with an M60 Machine Gun.

We finally arrived at Fort Stewart, Georgia, close to midnight; that was one long bus ride. It was close to 11 hours. Later, I found out that it takes a little less than four hours to travel from Fort Jackson, South Carolina, to Fort Stewart, Georgia, under normal circumstances.

A lot of things happened during the in-processing at Fort Stewart. But the one that sticks out the most in my mind was that when I was finally sent to my unit, they were getting ready for an Inspector General (IG) inspection. An IG inspection is (in short) an external group inspection that ensures a company is doing what it is required to do to be combat-ready and reports its findings to the Department of the Army at the Pentagon. Part of the inspection is to check to ensure everyone in the unit has a military driver's license. I never drove a vehicle in my life and did not have a civilian driver's license either. To make sure they were at 100% during the inspection, my unit's leadership gave me a military driver's license. Because of this, I asked other Soldiers to help me get a civilian driver's license. I met other Spanish-speaking Soldiers, who showed me where to go in Hinesville, the town outside the base, to get my driver's license.

Finally, I began to work as a personnel records clerk. My first task was to review the files of the soldiers stationed at Fort Stewart. I was working with other clerks. We had to fill out a prepared memorandum and check the blocks of required documents that were missing. At the bottom was the signature block of our supervisor. Once we completed a file review, we had to send a memorandum, checking missing items or identifying documents that needed to be updated.

Now, our supervisor had to sign this memorandum. As I completed a file, I took the file with the memorandum for my supervisor to sign. After a while, he told me that I could sign for him. So I started to sign *his* name. I must have done this well over a hundred times before someone saw me doing it and asked me what I was doing. I told him that the boss said that I could sign for him, so I signed for him. My coworkers called the boss, who asked me why I was doing that. The same supervisor who told me that I could sign for him said it was illegal and I could get punished for doing it. I reminded him that he said I could sign for him. My supervisor and coworkers all had to explain that signing for somebody meant to add the word "for" next to the supervisor's name, then sign your name over his/her name on the signature block. I was about to lose rank before having any. Thank God these were not documents that were that critical, or I would have been in big trouble. That was an interesting learning experience that could have ended my military career as a personnel records clerk as soon as it started.

A few months later, I decided to go to the Hinesville, GA DMV and apply for a drivers' license. While I was in line with all my documents, a policeman was going through the line making sure we had the required documents. When he got to me, he saw my military license and military ID and told me to go to the other line. He was a cop, so I followed his orders. When I got to the front, the lady asked for my documents, took my picture, gave me a paper showing that I would have a driver's license mailed to me in a few weeks, and that was that. No written or road test, and now I had a driver's license. CAN YOU BELIEVE THAT? I had never in my life, up to that point, been behind the steering wheel of any vehicle and was now in possession of a legit driver's license authorizing me to drive.

I considered it my luck (insert sarcasm here) that they gave me a military license during my in-processing that resulted in my name being entered in the duty driver's roster. But imagine my surprise when I saw my name on the duty roster to be the driver to the duty officer of the day. Lucky for me, I had a couple of weeks to learn how to drive. A friend of mine took me to a parking lot after work a few times and gave me a general idea of what I needed to do, and he let me practice in his car. I am a fast learner who pays a lot of attention. His instruction was good enough, and after a few days of practice after work, I felt ready to perform my duties as a duty driver. Thank goodness I did not have to leave the base to deal with a lot of traffic, and most of the driving was done after 9:00 PM. By then, most people in the Army base had settled down, and there were not too many people driving.

Nine PM came faster than I was expecting. Off we went all over Fort Stewart as the duty officer checked buildings and other units' CQs to ensure the buildings were securely locked and those on duty were doing what they were supposed to do. Not everything went well. There I was. I dropped off the officer in front of a building. As he went into the building, I thought everything was perfect. When I attempted to drive in reverse, I drove the utility truck into a light-pole. I was really lucky! There was not anyone around and no damage to the light-pole or the truck. I did not report the incident to the staff duty officer. I certainly was scared to death. Subsequently, we drove to a location probably five miles away on one of Fort Stewart's many back roads.

The truck's dashboard was dim. Due to my inexperience driving, I thought it was normal. I could not see how fast or slow I was going. I was probably driving slower than Hoke Colburn from the movie *Driving Miss Daisy*, but I did not want to get in trouble with the officer. So I kept driving slowly. I had no idea that there was a knob to make the dashboard bright or dim. I remember the officer staring at me and probably thinking, *What the hell is wrong with this boy?* We got to our destination about 45 minutes later—a trip that perhaps could have taken 15 minutes. Soon after we returned to the headquarters, the duty officer decided to get the keys to the vehicle and leave without me a couple of times after that. I did not drive the rest of the night, nor the rest of my time at Fort Stewart, which was okay with me!

The next day I told my friend what happened, and he could not stop laughing. He then showed me the knob in the vehicle and what to do the next time. He said, *"Well, at least you did not have to drive the rest of the night!"* Well, the duty officer probably drove a lot faster without me slowing him down!

Baseball, baseball—I love baseball! One day as I was throwing a baseball up in the air near the barracks, a Mexican-American Soldier, who I had met earlier, came with his glove and asked if we could play catch. We played for a while. He asked me if I was playing with a team. I told him that the last time I played was in Puerto Rico. He invited me to practice with his team off-post. I found out later that the team played in the Negro

Baseball Minor League in Georgia. I practiced with the team; the coach saw my talents and asked my friend to ask me if I wanted to join the team. I did; baseball is my favorite game!

Soon after that, we went on the road on a Sunday afternoon for my first game. The team bus took us to a location deep in the countryside of Georgia. I was ready to play ball! Being the new guy on the team, I did not start. I must add that these guys were impressive baseball players, and as skilled as I was, if not better, sometimes I felt like I was playing amateur baseball, and they were in the Major Leagues. The coach finally put me in the game as a pinch-runner in the third or fourth inning. As soon as I stepped onto the first base bag, the other team came off the field. I did not understand what was going on. So I asked the first base coach in my limited English what was going on. He did not know either or did not understand me. The next thing I know, my manager, with my Mexican-American friend and the other team's manager in tow, were walking in my direction. The other team's manager asked me a couple of questions that I did not understand. *My friend translated some of the conversations.* Then the manager told me to say something; when he heard me talk, he said it was okay and that I could play. I asked my friend what was going on, and he told me that the other team (being a Negro

As soon as I stepped onto the first base bag, the other team came off the field.

League) did not want to play against a white guy. I asked him who was the white guy. And he told me, "YOU pendejo, they are talking about YOU!" I was shocked that someone would consider me white, but at the time, I was not as dark as I am today. Yes, you guessed it, I thought it was funny, and to this day, I still laugh about it.

I was frustrated. It was time to learn English. I said to myself; you are not going to learn English when you continue to speak Spanish during your free time.

Baseball been bery, bery, good to me. LOL

I was determined to do something. I was already spending about an hour before bedtime with one of my roommates, learning the basics. My roommate had a dictionary, and almost every day, he had me look up an English word, learn the definition, and try to make a sentence or use it in casual conversation. But I still felt like I needed to do more. After all, I worked in an office environment, and it was hard to communicate with others. Good thing that my supervisor sat me next to a Mexican-American, which helped me a lot. I know that sometimes I got on his last nerve because I have an inquiring mind and ask many questions. Nevertheless, again, I felt like I needed to do more to speak the language better.

After thinking about it a lot, I decided that the best way for me to learn would be by not speaking Spanish as much, so I did not spend a lot of time with my fellow Puerto Ricans. After a while, a few of them came to visit me in the barracks to ask me what was going on. I told them that I had been busy. They started calling me names ("traitor" was just the nicest insult). They began to say that I sold my country and chose not to speak to me when they saw me. My compatriots began spreading this malicious gossip. It was a good thing that those who were older knew better and took the time to communicate with me. I told them what I was trying to do. They perfectly understood because they had done the same thing.

I still do not understand, up to this day, why someone going to another country would not try to learn the native language and the customs. I did the same thing later in my career when I was stationed in Germany and South Korea. German (Deutsch) was a lot easier than Korean (Hangul), but I picked up a little bit of both to help me break the ice with people from those countries.

I can say that I was blessed with the two roommates I had in Fort Stewart, Georgia. One was white, and the other African-American. I learned a lot from them. The white roommate spent a lot of time teaching me how to speak English, and we had many conversations about everything. He was from Yakima, Washington. My other roommate was from somewhere near Columbia, South Carolina. At that time, I could not understand why their English was different.

Another one of the "first-in-my-life" experiences was also at Fort Stewart. I did not have a television in my room because, quite frankly, I could not afford one, and neither could my roommates. Soon I found out that the Army, and probably all the armed services, have this building called Recreation Center or "Rec Center." This is a place where the single Soldiers living in the barracks can go for entertainment. They have televisions, pool tables, table games, arcade games, books to read, snacks, and the best part is that everything is free. One day, when I got to the Rec Center, I saw this strange game on the television. They were throwing a big egg-shaped ball and tackling each other. Then I saw this guy running with the ball with two guys hanging on his back and trying to tackle him with no luck. He was dragging them with him. The name of the game was American Football. The guy's name was Herschel Walker. Imagine that—the first time I watched a football game was on television at the Rec Center, and the great Herschel Walker provided the most memorable experience of the game.

I will never understand why people classify as stupid those who do not speak their language and can be cruel to them. That is exactly how I felt, *stupid*, and what they called me because I did not speak English. "You're nothing but a non-English speaking mother f***er who will not last four years in the Army." I heard this many times as I swept the sidewalk of the legendary AG Row at Fort Stewart. At the time, AG Row was a line of buildings with personnel, transportation, finance, and many other operations. These approximately 20 buildings supported the base's mission and were connected by a long off-the-ground walkway. That was my task, without any help, for about six months. Yet, it felt like an eternity. No matter what uniform I was wearing, my routine did not change: Make the first work formation (to be counted as present for duty), move to AG Row, and start sweeping because it would take me between two to three hours to do this by myself.

> *"You're nothing but a non-English speaking mother f***er who will not last four years in the Army."*

I cannot tell you how many times my coworkers, some with the same rank, walked by me and laughed. Some of them had not been in the Army for as long as I. Not once did they have to sweep the sidewalk. I did not understand many of their comments but saw them laugh plenty of times as I swept all by myself. But no matter what, I did the best I could and was proud of my work.

I would see a Sergeant Major by the last name of Shealy every day. I always said good morning to him, and he said it back to me. A few weeks into our daily exchanges, he called me over to his office. As you can imagine, I was astonished. All kinds of thoughts crossed my mind. Already in his office, I felt at ease when he asked if I wanted to get promoted to sergeant.

I replied, "yes." However, at the time, I only had "mosquito wings" (the name we used to make fun of the rank of Private Two, which resembles mosquito wings) on my shirt collar. He told me that I had to think about my future. The Sergeant Major helped me by signing me up for Army Correspondence Courses. After finishing each course, I came back to him so that he could sign me up for another correspondence course. They were easy to do because I could read English; I only had trouble speaking the language.

AG Row. It looks like it goes forever, and many times as I swept it by myself, it felt like it did.

As I mentioned, people can be cruel when dealing with others who do not speak their language, and, of course, they have the upper hand. On one occasion, a female coworker was taking lunch orders; she was going off-post instead of eating in the dining facility. She asked me if I liked fur burgers and if I wanted to order one. I wondered what she meant, and she kept asking me if I wanted one while others in the section laughed. I asked Go-Go, the moniker we called Gomez, the Mexican-American Soldier I sat next to at work, what she meant. He is also the first guy

I saw that could type a lot faster with two fingers than I and others could do with ten. Anyways, he told me that she was messing with me, but he would not tell me what it meant. I told him that either he translated what she was saying, or I was going to ask someone else. Go-Go cared for me and told me before I got in trouble by asking the wrong person. A fur burger is the slang word for the female body part between her legs. Imagine my shock when he told me what it meant! The next time the other female coworker came around taking lunch orders, I told her that I wanted a fur burger but not hers. She blushed and never asked me anything else, and stayed away from me after that.

On another occasion, my coworkers fabricated Permanent Change of Station (PCS) orders, sending me to Germany, thinking that I would be upset. Little did they know that I joined the Army to get out of Puerto Rico and travel. They had it all meticulously planned to the point that the orders looked like actual orders. They probably did it to scare me since I was still not particularly good at speaking English and was now, according to their joke, going to a place where they spoke a different language, German, a language that I also did not speak. They came to me with sad looks in their eyes and told me that they had to give me some bad news. That is when they gave me a copy of their bogus orders sending me to Germany. It backfired on them or worked, depending on how you look at it when I saw the orders and got overly excited about going to Germany.

That country has been in my dreams since childhood, one of the places I wanted to visit. They eventually had to tell me that the orders were not real. I could see the disappointment in their eyes when I was not scared by their prank. The bad news about that experience? I was disappointed that I was not going to Germany. The good news? That changed. Read on.

A few months after that joke, I was on orders to Germany, and this time it was for real! Maybe they felt so bad about their prank and how excited I was to go to Germany that they called their contacts and made it happen. With the orders to Germany came the opportunity to attend a German Headstart Class at the Ft. Stewart's Education Center. We learned German customs, practiced for taking the international driver's license exam, and—after all—even got to learn a little bit of the German language from the teacher, who was a German-American married to a Soldier. And so I was off to my next journey. Fort Stewart was not a "permanent party" at all.

A few months after that joke, I was on orders to Germany, and this time it was for real!

In December of 1981, after leaving Ft. Stewart and spending a couple of weeks on vacation in Puerto Rico, I got on a plane on my way to the port of call section at Charleston Air Force Base in South Carolina. Port of call is a technical term used in all official shipping documents. For the military, a port of call is where military members and their dependents moving to an overseas location are processed.

Now, it happened again. Another "funny" experience. This one occurred while I was waiting to board the plane to Germany. Since I was not flying until the next day, I spent the night at a motel right outside the airport.

The following day, I decided to eat at the restaurant close to the motel before reporting to the airport. I saw hash browns on menus for quite some time but did not know what they were. The restaurant had hash browns on the menu, so I ordered some. They were quite good. Now, the name, hash browns, intrigued me. Why do Americans call this dish "hash brown"? I was asking myself. After all, I saw potatoes. But at least now, I had learned what it was. What was funny about the hash browns? Stay tuned for the funny part. I then reported to the Air Force Base airport and was soon on my way to Germany.

Garlstedt, Germany

I ARRIVED AT THE RHEIN-MAIN Airbase, which was next to the Frankfurt International Airport in Germany on 28 December 1981. We were transported to the Replacement Detachment, where we spent a couple of days processing before going to our respective units in the country. One of the personnel specialists, who saw my occupational specialty on the military orders, pulled me to the side and told me to remove my old unit patches from my uniforms. He told me about the "crazy" Post Command Sergeant Major waiting for us at our destination that would look for any unit patch other than the one representing them and go off on people—regardless of their rank. I hesitantly took the patch off, and I trusted that he gave me the right advice since he was very insistent. He kept coming back to me during the processing to tell me that I would thank him later.

A couple of days later, I was on a train going north to Bremen, Germany, and from there, to catch a bus to my destination of Lucius D. Clay Kaserne in Garlstedt. It was a long-distance, so we got sleeping berths on the train. I slept for most of the ride; it was the red-eye train. I did not sleep too soundly for fear of missing the train stop.

When I heard the name of the town where I was supposed to exit (Bremen), I jumped out of the sleeping berth, got my stuff, and left the train. A bunch of us got on a military bus that took us to the base. In my haste, I forgot that I placed my wallet under the pillow on the train.

My heart dropped! I worried and worried for weeks. A month later, the unimaginable happened! My wallet was returned to me *with everything in it*, including the money! Yes!, Unbelievable and incredible! Thank you, Lord!

Not too long after that, we were at our destination of Lucius D. Clay Kaserne—which was home to the U.S. Army's 2nd Armored Division (Forward) at the time. The Kaserne (barracks) was located in Garlstedt, between the towns of Bremen (22 km south from our location) and Bremerhaven (39 km north from our barracks). Coming through the gate was when I first noticed that I did not have my wallet. The bus then took us to a parking lot across the street from the post headquarters. And who was waiting there? You guessed it. The crazy Post Command Sergeant Major that my fellow personnel guy warned me about in Frankfurt.

While I was still sitting inside the bus, I could hear him screaming at everyone as they were coming out. He ripped the patches off quite a few uniforms regardless of their rank, officers included. When he got to me, he could not find the patch, took a look at me, and told me, "You must be an AG Puke, and one of your buddies in Frankfurt told you about me. GET THE HELL OFF MY BUS, and don't look back." I was smiling after that and wished I had the phone number of my counterpart in Frankfurt to call and thank him.

Since I arrived at my unit on 31 December 1981, there were not too many people there. I remember that the base was covered by snow, giving it a beautiful post-card appearance. The supply sergeant gave me some linen, issued a meal card, gave me the keys to a room, and told me where it was. I was to report for a briefing after the holidays. I thought I was in a room by myself, but some guys showed up a couple of days later.

They were not too receptive to me until they got to know me better. A few days later, they were smoking something that smelled strange in addition to marijuana. The Army was not too strict about those things at the time. In those days, the Armed Forces had no drug-testing systems. As a result, the use and abuse of marijuana and other narcotics were prevalent and detrimental to good order and discipline.

As these guys were smoking, they offered me some *hashish*. However, I thought they were saying something else. Yes, you guessed it again. I thought they said, "hash browns." Imagine that! In my ignorance, I asked them why they were smoking potatoes. They thought it was funny. But I felt stupid. I had never heard about hashish, and here I was, confusing one thing with the other. And no, I did not try smoking hashish; never been into drugs or alcohol.

I watched the Super Bowl with them. I still did not know too much about the game but, I vividly remember that Joe Montana was on fire, and the San Francisco 49ers defeated the Cincinnati Bengals for their first-ever Super Bowl victory.

Compared to Fort Stewart, Germany seemed to be the place where the permanent party label fit what the name implied. It seemed like that was what everyone wanted to do. But imagine my surprise when after almost two years in the Army, I was introduced to a GI Party.

I remember the platoon sergeant telling us after a Physical Training (PT) formation about the GI Party that night at 1800 hours (6:00 PM) in the barracks. I was looking forward to it. I got off work, devoured dinner, went to my room in the barracks, showered, and dressed up for the party. Sad to say, I found out that it was not that kind of party. A GI Party is a thorough cleaning of the military barracks by every unit member, including those who did not live in the barracks. Yep, I was surprised too. How could they spoil such a lovely name by associating it with cleaning/work? This was a party where everything had to be spic and span, and it lasted until almost 2200 hours (10:00 PM). It was really an intensive, top-down, inside-out group cleaning of the barracks and supervised by the platoon sergeants under the First Sergeant's (the top sergeant in the company) direction. So much for a party-party. After this party, we did not have enough energy for any other party.

However, people did party a lot. Some of them could not wait until after work, and on Fridays, they drank at work. I never did care for any alcohol. On Friday afternoons, a couple of the people I worked for - or with - would have me split a Coke with them. They bought the Coke; I drank half and gave them the other half. Little did I know that they were filling the other half of the Coke can with rum, and they were drinking rum and Coke at work. How ironic that, in typical Army fashion, they were "camouflaging" their drinking! Most of them were half-drunk by 2:00 PM. Like I said before, during those days, the Army was not too strict on drugs and alcohol or even fitness. The rules were too lax, but they would soon tighten up.

A mentioned, baseball was my game, and soon enough, softball became my game. In the Continental United States, I could play baseball. That was not an option in Germany at the time. All they had was softball, and I would not say I liked it. Nevertheless, I had to adjust and play that game. Compared to baseball, softball was relatively easy. It is not an easy sport, but it is at least to me when compared to baseball.

I ended up getting selected for the post team. So I traveled all over Germany, playing against teams from other military bases. Since we were in Northern Germany, we had to travel south on Thursdays. It took between nine and twelve hours, depending on our destination, to be ready to play on Fridays.

We spent the rest of our time, at the base, practicing softball. Yep. It became our full-time job. Imagine that: I was in the Army, but my job was to play softball. However, I have to mention that my bosses were not too happy with this "job."

> *The chief told me that he just got a phone call from the general—yes, the big guy—wanting to know why his shortstop was not at practice.*

The day after the tryouts, I went to work and told my boss that I had made the Post Team. He said that it was good, but I was not leaving work to play softball. So I did what every Soldier would do. I stayed at work when I was supposed to be at softball practice as directed by my boss. After practice started, my coach called the office about an hour later, asking why I was not at practice. I told him that my supervisor would not release me to practice during the day; I only had permission to attend softball practice after work. He told me that he was going to make a phone call and to be ready. Thirty minutes later, my chief and the NCOIC (the sergeant in charge) called me into their office with my immediate supervisor. I thought I was in trouble. The chief told me that he just got a phone call from the general—yes, the big guy—wanting to know why his shortstop was not at practice. I shared with him what my immediate supervisor said to me. His reply? He indicated that my place of duty would be defined by my coach for the next nine months. And, so, I was on my way to enjoy playing softball for a while.

Although it was fun, I only played with the Post Team that first year. I did not like traveling for hours every week and did not feel right about playing softball when I was supposed to be in the Army. Plus, my responsibilities at work and with the Army unit as a whole, increased during this time. The coach allowed me to play with them when they were playing in the local area. I did play with the company softball team and found myself traveling south to Pirmasens, Germany, for the Commander's Cup Tournament.

Another unexpected and stupefying experience. There I was in a room with at least 20 individuals—all of higher rank—and suddenly, I could hear someone cussing in Spanish. This person was using the nastiest words that you could say. I looked around the room and did not see any Hispanics. I started to look at the last names on everyone's shirt and did not see a Spanish last name. I thought to myself, *I am getting so good at English that it is starting to sound like Spanish!* Because everyone was speaking at the same time, I could not pick who was cussing.

Yet the foul language in Spanish continued until I saw this blonde-haired, blue-eyed white guy with the last name Walker, and I asked him where he learned to speak Spanish. He told me in his native Puerto Rico. Talk about a countryman—he was from the mountains in Puerto Rico and, like me, never left the island until he joined the Army. His father was an American Soldier stationed in Puerto Rico. He married a Puerto Rican woman and liked the island so much that he decided to stay after he retired from the Army. The Walker I just met was one of their sons.

To a degree, Walker inspired me to improve my English further. Ironically, my English took a turn for the better as I was watching a movie. The main character was stealing a Russian plane, and the guy who was teaching him to speak Russian told him repeatedly to think in Russian. He stole the plane, and while up in the air, he had to read the instruments panel written in Russian, and he could not understand it because he was thinking in English and translating to Russian. He then remembered what the teacher kept telling him, think in Russian. Finally, he understood the instruments panel.

Eureka! I get it now—I said to myself. That was the same thing I was doing: I was thinking in Spanish. It was a key part of my problem. I was unnecessarily complicating my verbal communication in English since I was translating words, phrases, and sentences from English to Spanish and vice versa as I was communicating with others.

It constituted a milestone along the painstaking path to clear verbal communication in my second language. It enabled me to improve my English exponentially. Can you believe that? A movie revealed for me the key to effective oral communication.

Since my major shortcomings were in the area of oral communication, I ***had*** to rely on the written word—manuals, handbooks, memos, and written regulations to perform my job (just as I did in Basic Training). Eventually, I became a subject matter expert—I could find the answers to difficult questions and situations with ease. Some people did not take it well. Apparently, they were resentful or felt some envy. The fact of the matter was that I was not trying to demonstrate how much I knew or that I was the smartest man in the office. Since I could not communicate orally, I spent a lot of my time reading. That was my way of surviving and doing my job in the best possible way.

Here is another story about how reading paid dividends for me. I won many guard mounts—guard mount is a pre-shift official formation among designated United States Army Soldiers before performing 24-hour guard duty. This formation is usually held after all security personnel are "armed up" or received their weapons and equipment for the shift. During the formation, the Officer of the Day or the Sergeant of the Guard will discuss several topics, such as the current unit situation, current terrorism warnings or military intelligence, weapons, vehicle status, and ground safety. Members also will be inspected and assessed on their appearance, uniform, and equipment. A major part of the inspection was a question-and-answer session. The guard who performed best during the inspection did not have to pull guard duty. He was called "the supernumerary"—in essence, a substitute. The supernumerary gets the night off unless he has to replace another guard in an emergency. The supernumerary also got rewarded with the next day off and a certificate from the Battalion Commander. I won the supernumerary many times because all I did was read regulations and manuals.

While at work, I used to go toe to toe with my supervisor at the time, SP5 J.T. Williams. He was also deeply knowledgeable in our field, but I won a couple of bets against him on how to do things by knowing how to find the correct answer in books. During an argument, I pinpointed the exact answer in the Army regulation manual. Many times I had the Army Regulation and Merrian-Webster dictionary side-by-side. I was not familiar with many words and had to find their definition to understand what I was reading.

I told him that he should be ashamed of himself. He asked me, ashamed for what? I replied that I had been speaking English for a little over three years and could read and understand it better than him. Let me clarify; this was all in jest. He was one of my favorite mentors throughout my career; we consistently had a friendly relationship. However, the rest of the crew did not like that he spent a lot of time grooming me for success; apparently, they felt left out. Well....after all, I was the only one asking a lot of questions. For the most part, they did not show much interest in doing the job the right way. I remember that he used to call me "Cold Steel Cobra."

I was in for a surprise. For whatever reason, coworkers and supervisors kept a secret from this Cold Steel Cobra that he would attend the next month's promotion board.

They told another coworker that we were going to the board the next month and to keep it a secret from me. My coworkers finally told me at 5:00 PM the day before the promotion board. Why? What was the intent? Were they doing their best to make me fail?

I was frightened! I felt very cold like I was freezing; however, at that same time, I was dripping with sweat.

Well, it backfired on them. While I was studying that night, an Alert was called; we had an hour to report to work. An Alert is part of the pre-deployment phase of preparing for war. During this phase, a military unit will be alerted for possible deployment and could receive orders to mobilize. However, this was one of many done for training purposes only. We were at work until a little after 2:00 AM. Because of that, the promotion board was moved up a day. That gave me more than enough time to study since I knew most of the material, thanks to the uncountable hours studying for guard mount.

A good thing about the promotion board was that all the promotion board members knew me because I played for the post softball team. They asked me some situational questions that made me think hard, but, for the most part, I knew the answers from studying for the guard mounts.

Despite my knowledge, I was frightened! I felt very cold like I was freezing; however, at that same time, I was dripping with sweat. One board member covered his uniform's side that displayed the "US ARMY" tag while leaving his name tag unveiled. He then asked me for his last name. I was so nervous that, although it was clearly displayed, I could not see it!

The board members also asked me other crazy questions, such as how many inches were in a 30-inch step. They helped me along the way because of my language barrier; they knew I was trying really hard. After all the pains and tribulations, I did very well; the score reflected it. The members asked me how long I had been studying for the promotion board. They were shocked to find out I only had two days to study because of the alert, as I only found out about the promotion board two days earlier. They asked my coworker the same question. He replied that he had a whole month to study. The board members were not too happy when they found out that we worked in the same office. On top of everything, the Command Sergeant

Major called my chain of command, looking for answers on why I was not informed about the promotion board well in advance. Apparently, he let most of them have it good because they all apologized to me.

Not too long after that, the company found itself without a First Sergeant (1SG)—as mentioned, this is the key sergeant in any company. He did his time overseas and had to return to the United States. There was a Master Sergeant (same pay grade, but with very different responsibilities) at the Engineering Battalion who was supposed to be one of their 1SGs; however, there were no openings in the Engineering units. With a lot of hesitation, our battalion Command Sergeant Major interviewed and hired him as our new 1SG.

Talk about a cultural shock! Our outfit was an administrative one; accordingly, we had an administrative culture. The engineers, however, have a "ready to fight" combat mentality. What a contrast! We were not prepared for what we were about to receive. As an administrative unit, we did the Soldier thing; after all, we were in the Army—we just did not do Army to the extent and level of discipline that the combat arms units did. 1SG Booker came with a mission to "search and destroy," and I could not have been happier! At the time, I believed that we totally lacked the skills and leadership needed to survive in combat because of the lessons I learned in the "ready to fight" Army Division at Fort Stewart, Georgia. But that was about to change quickly, pronto, and in a hurry!

In less than a month, our entire noncommissioned corps (NCO corps or sergeants) were doing what they should have been doing all along. I did not have the rank to be a part of that group, but the 1SG saw what I did as a Soldier and treated me like one. Unexpectedly, 1SG Booker called me in front of the formation to inform the rest of the company that I was now in charge of all physical training activities. I soon became the company physical training NCO, even though I did not have the rank to qualify as an NCO. Can you believe it?

There are always good and bad people in every story—and then there was "Gringo," the best roommate ever! Of course, Gringo is not his real name. Gringo is a very humble and private individual that chose to remain anonymous. When Gringo arrived in Germany, we took to each other and became not only coworkers but good friends and soon roommates. I could always count on Gringo for good, honest advice. He was not that much older than me, but he was a college graduate and a man with a plan. You could say that Gringo was the perfect Soldier with an impeccable track record. He was a Godsend, and I learned so much from him. We have stayed connected throughout the years. He also can make the most "out of this world" spaghetti sauce that I have ever tasted. I asked him for the recipe back in the 80s, and to this day, almost 40 years later, I have yet to receive it!

During those days, it was a requirement to take an annual MOS (Military Occupational Specialty) Skill Qualification

Test (SQT). Based on the score, you could raise or lower your promotion points. Except for Gringo and me, everybody else in the section failed the test. In addition, somehow, the most recent SQT scores were missing from everyone's files except for—you guessed it—Gringo's and mine. Hmm....maybe it was because we were the only two who passed. We were called into the Chief's and NCOIC's (noncommissioned officer in charge) office. They told us what happened and asked us what we wanted to do with this information. They could all get disciplined if we wanted it, but Gringo and I decided that getting promoted before them was punishment enough, even though they tried to hurt our promotion chances.

We also told them to have the test results reprinted and place them in their promotion packets; they lost many promotion points. Some of the coworkers were scheduled to get promoted, but when the test failure was factored in, it reduced their number of promotion points. As a result, they no longer met the cutoff score. After that incident, Gringo and I got promoted because we met the cutoff score for that month. Gringo had a college degree and other correspondence courses that resulted in more promotion points. I relied on the promotion points that I got from all of the correspondence courses that SGM Shealy signed me up for during my first duty assignment at Fort Stewart. SGM Shealy retired, and I lost touch with him. It would have made him proud to know that he was responsible for my promotion to SP5. The other guy who went to the promotion board with me, the one my coworkers tried to help, did not get promoted until almost a year later.

With the promotion came more responsibilities. Luckily for Gringo and me (being sarcastic here), it came with a move to take over the EER (Enlisted Evaluation Report) Section. Gringo led the charge, and I was his student/coworker; it was a two men show.

Let me explain; an EER is a very important document. It has a high impact on promotions, assignments, and other personnel actions. Actions that are decided by people who do not know the individual Soldier and make decisions based on records. It is probably the single most important document a Soldier can encounter in their military career as it affects their future in the Army.

I distinctly remember our coworkers smiling and glad that we moved out of the section because of the SQT scores situation, although it was something that they created. They had a look of disdain except for a couple of guys that were on our side. At the time, the EER Section was in shambles and had the worst processing and timeliness rate Army-wide reported by personnel systems. It was an uphill battle. We did a lot of work and spent many after-work hours fixing the mess that was now for us to correct. We turned the mess into a flawless operation. As a result, the EER rates went from the worse to number one Army-wide. This situation made the Post Commander, Brigadier General Thomas H. Tait, so proud that he stopped his day and made the trip to our office to award us an Impact Army Commendation Medal. We offered to go to his office, but he insisted on coming to our location. An Impact award for outstanding achievement is intended to recognize a single specific act or accomplishment. It recognizes exceptional achievements and usually covers a short period with a definite beginning and ending dates (typically six months or less); in our case, we reached that mark in a little over 30 days but had to wait for the quarterly report to look at the results of our labor. You can imagine the look in our former coworkers as we received the award.

...the Post Commander, Brigadier General Thomas H. Tait, so proud that he stopped his day and made the trip to our office.

DEPARTMENT OF THE AR

THIS IS TO CERTIFY THAT THE SECRETARY OF THE ARMY HAS AW

THE ARMY COMMENDATION MED

TO Specialist Five Pablo A. Dias-Rodrigues, United States Army

FOR exceptionally meritorious achievement while assigned to Hea Headquarters Detachment/Adjutant General Company, 498th Support Batt October 1983 to 31 October 1984. During this period, Specialist Five served as Enlisted Evaluation Report Clerk. During the 3rd and 4th quar year 84, and 1st quarter fiscal year 85, he maintained an Enlisted Eva late rate of 1 percent and less. His late rate was the best Army wide except for 3rd quarter 84 and at that time his rate was second Army wid USAREUR. Specialist Five Dias-Rodrigues's outstanding performance keeping with the finest traditions of the military service and reflec upon himself, 2d Armored Division (Forward), and the United States Arm

GIVEN UNDER MY HAND IN THE CITY OF WASHINGTO
THIS 31ST DAY OF JANUARY 19 85

THOMAS H. TAIT
Brigadier General

It was so long ago that I could only find this copy of the award.

As individuals, sometimes we make dumb

comments, and sometimes the comments are so bad that we call them "whoppers" (a huge unwise or dumb remark). I made one whopper of a comment. Okay, if you know me, I have made many whopper comments. You could say that I have made so many whoppers that I could be confused with a cook at Burger King, but you get the point. Out of the many comments or gaffes I have made, this one is worth mentioning. For most Army units in Germany, there was a requirement to take advantage of the numerous historic sites by conducting a day trip to those locations. This one time, we went to a concentration camp.

I did not know what it was other than it was a cemetery. We rode the bus for over an hour and arrived at our destination. It was nice looking at the German countryside on the way to the camp. As soon as we exited the bus and before the guided tour began, I commented to Gringo "who put a cemetery in the middle of these hills?" That was my whopper of a comment. Gringo told me that those were not hills but instead thousands of bodies covered with dirt.

Welcome to Bergen-Belsen, the Nazi German concentration camp. This is the same place where Anne Frank died and was buried. Even though the concentration camp did not have a gas chamber, over 35,000 people died in Bergen-Belsen during a period of four months in 1945. They died from starvation, from being overworked, and from many catastrophic diseases. Talk about a whopper of a dumb comment, "who put a cemetery in the middle of these hills?"—I will never forget it. It brought a new attitude about my reason for being in Germany.

More moments that I will not forget. The new Post Command Sergeant Major knew 1SG Booker and decided to do PT with our Battalion and run with our company. The previous one, the one barking when we arrived, did his time and returned to the states. Soon after, we began to run, with the 1SG calling cadences and leading the way; he then gave me the company's reins for me to call cadence. Because the run's pace was faster than what we were accustomed to, many people were "dropping like flies" out of the run. We ran, and I sang for six miles, and nobody came out to help me sing. I thought that I was surely going to have a heart attack. After the run, the few who usually came to help me call cadences told me they could not help me because of the fast pace. They were trying hard not to fall off the run themselves.

After the run, the Post Command Sergeant Major spoke to the battalion to congratulate us for the excellent run. It is worth mentioning that he was not breathing hard and was acting as if he did not run at all, but I saw him run circles around the entire battalion formation that consisted of over 600 Soldiers. It was unbelievable! I was in my twenties at the time and was gasping for air. He was barely breathing hard. A few years later, he would become the eighth Sergeant Major of the Army; his name is SMA Julius W. Gates.

Speaking of Command Sergeants Major and US Army Sergeants in general, I learned from many of them how to do the very best in all tasks. Before the Army Sergeants, however, I had another "Soldier"—my dad. He raised us always to do the best we could do at everything, which I do to this day, to the chagrin of many. However, it is not my attempt to overshadow anyone. It was the way we were raised. He used to say, "Do everything as if you are doing it for the Lord." I took this to heart, even though I was not a practicing Christian.

Not bragging—okay, I am—but one of those things that I did better than anyone in my company was waxing the floor in our work building when I was on 24-hour duty. It helped keep me up and alert after midnight, and I loved seeing the end product. I always stripped the entire floor and mopped it a few times with clean water to remove any dust particles left before I waxed and buffered the floor. It was a tedious task, as you had to use a floor buffer to strip all of the old wax, then mop the floor several times, with clean water each time, to get rid of the stripping material because if you failed to remove all of the stripping material and laid wax as soon as you began to buffer the floor, you would get all sort of black lines on the floor. It would look like crap from the grime all over it.

I spent many hours doing this, mostly trying to stay awake because we were not supposed to sleep during duty. If we did fall asleep and got caught, it could have been the end of our Army careers. Most of the time, I started at midnight and ended at about 5:00 AM, right before people reported to relieve me so I could go to breakfast. I am proud to say that I could make that floor look like glass!

But because of this, every time an important person came to visit, I found myself on 24-hour duty. I did not think it was fair because a duty that we usually did twice a month (one weekday and one weekend) I was sometimes doing twice a week.

One of those times, because I was sick, I did not put forth the maximum effort as it was the norm and expected by my supervisors. The next morning, the Lieutenant chose not to release me for the day at 8:00 AM because I did not shine the floor like they were used to seeing. Nobody else did it as I did, yet I was being punished.

As he instructed, I soon grabbed a razor blade and began to strip old wax from the edges of the floor by the walls on the hallway. About an hour later, the AG (Adjutant General) Sergeant Major (SGM) came to me and asked what I was doing. He asked me what I did to get punished. I told him about being sick and the Lieutenant not liking that the floor did not look as it typically did when I was on duty. The SGM got really upset. He released me, instructing me to go to the barracks to go and get some rest. Before I left the building, I could hear him yelling at the Lieutenant as I had never heard anyone yell at another person before, including the drill sergeants. The Lieutenant went to complain to the Adjutant General (AG), an Army Major. The Major asked the SGM what happened. After the AG learned the whole story, he also lit up on the Lieutenant.

Later that day, after work, some of my coworkers came to my room in the barracks to tell me that the Major and the Sergeant Major had the Lieutenant strip and wax the floor and that it looked like crap. They also told me (laughing) that I would probably be the one to clean up the mess because there was another VIP coming to visit, and nobody could do the floor as I did. That made me feel good—not the Lieutenant getting punished, but that my coworkers complimented me!

The next day the Lieutenant apologized to me. He even brought me pizza the next time I was on 24-hour duty, maybe because I cleaned the mess he made. And after stripping and waxing the floor himself, he understood that it was not an easy task.

Germany was where I really learned how to drive. I drove a little bit at Fort Stewart but not to the extent as in Germany.

Ich Liebe die Autobahn; I love the Autobahn! Picture this, here I was on the Autobahn, flying at about 120 mph; at least I thought I was flying. I was paying attention to the road ahead of me while checking my side and rearview mirrors for other vehicles. I was enjoying the ride and thinking, "wow, I cannot believe how fast I am driving!" However, keep in mind that driving that fast only happens on the Autobahn; once you exit, there are normal speed limits for city driving like in the

U.S. Back to the story, I was flying, listening to my music, and admiring the speedometer in my brand-new black 1984 Nissan Pulsar NX. It was an unbelievable feeling as I could do this and not worry about getting pulled over by cops.

Then out of nowhere, I saw this thing, through my rear-view mirror, coming towards me faster than anything I have ever seen. It flashed by me so fast that its aerodynamics swayed my car. I could not even recognize the type of car it was; It was like a silver bullet flying by me. We both took the next exit, and only because there was a red light at the exit I was able to catch up to it. It was then that I could tell that it was a Lamborghini, and I said to myself, "Wow, what a nice toy!"

It flashed by me so fast that its aerodynamics swayed my car.

And absolutely yes; I wish I could drive one, did not want to own one (too expensive), just drive it on the Autobahn while screaming, "I HAVE A NEED FOR SPEEEEEEEEED!"

After all, I had an opportunity to drive a SAAB 900 Turbo with Gringo, and we had a blast. When we stepped on the gas pedal, the car sped off like a plane on a runway before lifting into the sky; what an incredible feeling!

Yes, it is very different when you are young, inexperienced, and unwise. I can reflect in retrospect and, therefore, recognize how immature and irresponsible I was. I am grateful because I did not have an accident. My perspective and sense of responsibility have changed as the "Cause of causes" has been teaching me through the "School of Life." The lectures and classroom teachings have been challenging but worth the pain.

The mighty Nissan Pulsar NX. I drove it at 120 MPH on the Autobahn.

I Do Not Need to Get a Job; I Am Still in the Army

IT WAS ALMOST THE summer of 1984, and after three and a half years in Germany, I finally returned to the United States. I was now going to physical therapy school at Fort Sam Houston in San Antonio, Texas, but not until I spent 45 days in Brooklyn, New York. My parents, who lived in Puerto Rico, moved there soon after I moved to Germany. Since I did not return to the States for three years, my vacation days accumulated. Thus, I could take that many days of vacation while in transit to my next assignment.

Not long after I arrived in Brooklyn, New York, and picked up my car in New Jersey, someone broke into it and stole my Blaupunkt stereo. Blaupunkt was a German manufacturer of electronics equipment, noted for its home and car audio equipment. It was one of the leading German manufacturers, and its products were expensive and a novelty at the time.

A few days later, the coward sent his pregnant girlfriend to try to sell it back to me.

A few days later, the coward sent his pregnant girlfriend to try to sell it back to me. The idiot had the nerve to have her ask me if I would buy it back for $750.00, which was the value of a new one. He sent his girlfriend to me because when he stole it from my car, the plug that connected the radio to the car came off, and he could not use the radio or sell it to someone else without it. After much negotiation, I told her that I would give him $100.00. She came back insisting on $700.00, and I told her that thanks but no thanks; I had filed an insurance claim, and he could do whatever he wanted to do with it. I did not need to pay $700.00 for something that was mine. I told her once again that he could have it and good luck selling it without the plug. It is worth noting that the plug connecting the stereo to the car consisted of a set of pins explicitly designed for each particular unit; it was a safety feature, and there were no two connections alike. It worked like a fingerprint that you could not duplicate. He could not go anywhere to buy one. Without it, the radio was worthless. The coward sent her back to me to ask if I would give him the plug. I took a hammer, broke the plug to the radio in front of her, and then threw what was left away. Good luck using or selling that one. By the way, that action was not indicative of my real nature. I thought, however, that the criminals needed a "benign" lesson.

Yep, he couldn't use or sell the stolen radio because the plug stayed in the car.

It was great; I finally saw my dad and mom again. For some strange reason, my dad thought that I was out of the Army and kept telling my mom that I needed to get a job. No matter how many times I told him that I was on vacation, he kept thinking I was lying. He kept telling my mom that I looked sick because I was pale; the average summer temperature at my location in Garlstedt, Germany, was around 60 degrees Fahrenheit. Yes, it was cold year-round, and the sun was not strong enough to help anyone get a tan. So I was pale when I returned from Germany. But not because I was sick. Now that I was in Brooklyn, New York, I finally got to work on my tan!

What am I? Black, White, or Puerto Rican?

AFTER MY LONG VACATION, I was off to my next assignment in San Antonio, Texas. Fort Sam Houston and physical therapy school, here I come! I had never driven that long and that far in my life. The trip from Brooklyn, New York, to San Antonio, Texas took about two and a half days because it was just me; it might have taken me longer if I had a family.

During the trip, I remember stopping at a restaurant near Roanoke, Virginia, to get something to eat. I was stopped by the waitress as soon as I entered the establishment. She told me, "We don't serve N*****s here." There is that word again, but this time I knew its meaning. Initially, I froze in place, looked behind me, and asked her if

*"We don't serve N*****s here."*

she was talking to me. Then I told her that I was not black; I was Puerto Rican. She then told me to come on in. To this day, I still do not understand why it was okay for me to eat at that place, but they would not allow African-Americans. I told her that it was okay and that I would eat somewhere else and left. She yelled for me to return, but I kept going. Many years later, while assigned to a location a few hours away, a coworker and I went to a fellow Soldier's funeral, we drove by this place, and I told him the story. He, being funny, asked if I wanted to stop there for lunch. I declined and told him that although it has been several years, maybe 18 years, I had no desire to relive the moment.

I remember stopping at a gas station, eating snacks instead, and asking myself, "What am I? A few years back, I could not play baseball in the Georgia Negro Baseball League because I was white, and now I cannot eat because I am black. Dang, all this time, I thought I was Puerto Rican." Thank God, the rest of my trip to San Antonio was uneventful.

My first day at Fort Sam Houston in San Antonio, Texas, was in the middle of the summer of 1984. One of the cadres at the Academy of Health Sciences was a fellow Puerto Rican Sergeant First Class (SFC). Even though he had been in the Army for 16-plus years, we could not understand what he said every time he spoke. Little did I know that English was not the only language he had trouble speaking. He also did not speak clear Spanish either! Anyway, this one time, before beginning our physical therapy school, we had to do some housecleaning, or area beautification, as we call it in the Army. The SFC was dividing the work amongst all of us, approximately 80 Soldiers. My fellow Soldiers were asking me to translate even though he was speaking English. At this time, my English was still somewhat limited, but I knew that it was a lot better than the SFC's. He told the entire group that volunteers with driver's licenses were needed and to raise their hands if they had one. By this time, I had been in the Army long enough to know not to volunteer for anything, also because he looked at me with a smirk.

Although I had a license, I did not raise my hand. Hands went up quickly, as they probably expected to be driving a truck somewhere and doing the easy work. About 12 raised their hands. The SFC told them to leave the formation and step to the side. As soon as they were all out of the formation, he asked them to show him their licenses. He was grinning at the rest of us. Soon after he verified that they had driver's licenses, he told them to go over to the push-mowers (as he was pointing at them) and start driving them all over the company area until they mowed all the grass. They thought they were getting an easy detail; instead, they were pushing lawnmowers the rest of the day. I was laughing so hard I was in tears!

Here I was, finally, at the Basic Medical Orientation Course. It was a requirement before attending the Physical Therapist Course; my reason for being at Fort Sam Houston. It was an exciting course. A few of us had already learned most of the information during our regular Army duties. It was mostly a glorified First Aid course. One of the very few things in this course that none of us ever did was insert a needle into someone's arm to connect an IV. I saw a couple of my classmates passing out as soon as they drew some blood. I did not know how it related to physical therapy, but it must have been just teaching us basic medical knowledge before attending our medical courses. This was an uneventful course, and a month later, we all graduated, even those who passed out at the sight of blood. After the course, we all went in separate directions, as some were going to the pharmacist course, x-ray technician, laboratory technician, physical therapy course, and many other courses.

Physical Therapy School came with one of the more challenging situations I have never encountered. Here I was, recently promoted to Sergeant, and because of my rank, I was now the platoon sergeant to a group of medical students. Challenging is an understatement. Not only did I have to learn to lead at a level two ranks above my recently promoted rank, but I also had the duties and responsibilities of a student at a challenging course. Besides my platoon sergeant's regular

Physical Therapy School came with one of the more challenging situations I have never encountered.

duties, I had to keep up with all my classwork and my classmates on and off-duty, on top of my language handicap. Most days, I went from class to a meeting with the leadership, study hall, and then studying until 2:00 am to learn and retain the long medical terms/words. Words such as myofilaments, sarcoplasmic reticulum, tropomyosin, adenosine triphosphate, sternocleidomastoid, thoracolumbar fascia, and infraspinatus, well . . . you get the point. Imagine my conundrum. I had trouble saying sheets, and now I had to say and know the definitions of these parts of the human body. As if I already did not have enough on my language barrier. It was not easy, but I was determined to accomplish my mission in all areas.

During the course, I realized how evil individuals would use their positions to push their personal agendas. They would do anything within their power to destroy a person. For whatever reason, the teacher, an Army Captain, took some disliking to me. I was alerted by a couple of my classmates about the comments he made directed at me.

After taking a test that we answered using a pencil, as required by the instructor, I noticed that somebody changed some of my test answers. I could see where there were erasures on the test that I did not make. I began to pay more attention to my tests, noticing that it was reoccurring.

During one of our study halls, my classmates were unsure about a diagram, and I explained it to them. After the test, I remember that they came running to me because they earned a 100 on the test. I understood their joy because the diagram that I explained to them carried a lot of weight on the test. I had to tell them that somehow, I got that one wrong. They could not believe it. I could still notice that my answers were getting changed.

On the next test, I used a pen instead of a pencil. Somehow, I did very well on this one, which they called the most difficult. If I tell you that the instructor was pissed, it would be an understatement because he (by my assumption) could not change any of my answers. He sternly spoke with me about using a pen, and the next thing I know, he scheduled me for counseling with the Chief Instructor.

When I reported to the Chief Instructor, she made no mention of using a pen to answer the test. I did not mention it either. Instead, she told me that my instructor complained to her about my intimidating behavior. Shocked and in disbelief, I asked her to explain. She stated that when the Captain was talking to me about a class issue, I was eyeballing him with an intimidating look. I replied with something to the effect of my dad raised me to look at people in the eyes when they spoke to me. It was a sign of respect, showed that you were paying attention, and cared about what they were talking about. I also asked how could a recently promoted Sergeant intimidate a Captain. It did not add up. I guessed he was trying to intimidate me by referring me to the Chief Instructor. The

Chief Instructor spoke with me for a little bit and released me back to class.

However, the Captain was not done with me yet. When he saw that my grades were improving while I kept answering the test using a pen, he dropped me off the class two weeks before the end of the course. I would not go down without a fight, so I requested to see the Chief Instructor. I learned that the Captain misled her to believe that I had trouble understanding the written material due to my poor English. He assumed that I could not read. As a result, the Chief Instructor spoke with me, and upon her request, I went to the Education Center to take an English Comprehension Test. The test results were in my favor for apparent reasons and proved what I kept telling her: that I could read and write English, but something was going on with the Captain since he was unjustly punishing me. After the test results, I was still not allowed to return to finish the class. Consequently, I went to both the Inspector General and the Equal Opportunity Office to file a complaint.

The Inspector General did a quick investigation and told me that the Chief Instructor and the Captain were the course authorities. The Office of the Inspector General could not order them to let me finish, even though my grades were good enough and all I had left was the practical portion of the course.

The final answer was that, according to regulations, the course was for new Soldiers at the start of their careers; it was not open to those with the rank of Sergeant. I was selected for the course before my promotion to Sergeant. Thus, I did not qualify. Hmm—Why the authorities never notified me about this before? So that was the end of my formal complaint; I did not hear from the Equal Opportunity Office.

In the end, the Captain succeeded; however, as painful as this situation was, it worked out for me rank-wise. I would have retired with a much lower rank had I stayed in the medical field because of the lack of promotion opportunities. Nevertheless, I left the course with a deep scar and lack of trust in the "Army System." From that moment on, I had constant doubts about its fairness. I was also appalled by the deceitful and evil actions of certain Army "leaders." At any rate, off I was to my next adventure.

Once I left the schoolhouse, I found myself working in the Student Personnel Section. Somebody initially told me that I was on my way to another duty station, but the female NCOIC saw me and told the clerk they were keeping the cutie pie. And that is how I got to stay at Fort Sam Houston. Can you believe it? I never thought of myself as good-looking, but I was okay with that if it kept me there.

I got many opportunities at Fort Sam, many that I should have received while in Germany. Not long after I started working there, the NCOIC sent me to the Primary Leadership Development Course (PLDC) at Fort Hood, Texas. Upon my return, I was sent to the Basic Noncommissioned Officers Course (BNCOC) at Fort Harrison, Indiana.

PLDC was a great experience, and many of the tasks I learned there I continue to use to this day. I wish I had learned them before my time as a platoon sergeant at the medical school. This is where I learned the critical NCO tasks to make me successful throughout my career. My instructor was SFC Taylor. He was a great mentor and leader. He offered me the opportunity of getting stationed at Fort Hood to become an instructor, but I declined. I felt that it was not going to help my career with my rank at the time. I had other goals in mind, and Fort *Sham*—as many call Fort Sam Houston, seemed to be the right place to accomplish them.

Once I completed the training, I got to settle. It was the beginning of my personnel sergeant supervisor career! I oversaw 12 personnel Soldiers; eight happened to be females. Not that it mattered to me, but some people were wary that a new NCO like me would not be able to handle 12 individuals. I believe I did well, as my evaluations reflected.

I will never forget the time that six of the females were pregnant. One day I was told to report to the Branch NCOIC's office, where two other NCOs started asking me questions about their pregnancies. When I finally caught on to what they were saying, I asked them if I was under investigation. They answered that I was. I then asked them bluntly about what. They told me that there were suspicions that maybe I got my Soldiers pregnant because I was single and did not have a girlfriend at the time. I broke into laughter to the point that I was in tears. I told them that I did not do such things; they were crazy for thinking like that.

They told me that there were suspicions that maybe I got my Soldiers pregnant because I was single and did not have a girlfriend at the time.

The NCOs based their allegations on the fact that I was going to lunch with them. I told them that I went to lunch with the entire group and asked them if that made me gay since there were also men in the group. I kept laughing about their insinuations and told them once again that I did not do such things. My interrogation ended just as soon as it began. The NCOs then gave me a list of the six Soldiers to be interviewed. They also told me not to give them a heads-up. I sent the six Soldiers as the NCOs requested, and, one by one, they verified that I was not their expected baby's daddy. After that, it became a running joke with the entire section, but I did not think it was funny when they wanted me to buy diapers!

There was a lot to do in San Antonio and the vicinity. I will never forget the time I went tubing with a coworker at New Braunfels, Texas, not too far outside of San Antonio. My friend rode his motorcycle, and I drove my car and transported the cooler. We rented three tubes, one for each of us and another to hold our cooler with snacks and water for the approximately four-hour tubing expedition. We had a great time. When it was time to return to Fort Sam Houston, we decided that he would drive my car and me the motorcycle. Soon after we departed, it began to rain.

So here I was riding the bike in the rain on I-35, at its max speed of 80mph, passing cars left and right without a care in the world, IN THE RAIN!!! I am sure many of the people I passed on the highway prayed for me, or I would not be here today. When I finally got to my room in the barracks, my brain caught up to the rest of my body, and I realized that what I did was completely stupid and totally out of character. To this day, I cannot believe that I committed such a reckless act. My Lord! I put my life and others' in danger—no doubt! I called my friend, who resided in government quarters not too far from me, and told him to come and get his motorcycle. He told me to ride it to his house; I replied that I was not responsible or smart enough to drive a bike.

That was in the mid-eighties. And that was the last time I drove/rode on a motorcycle. This incident was not "lost in the translation;" it was "lost in my stupidity," and by the grace of a God, who I did not want to acknowledge at the time, I am here today.

Recruiting School

NOT LONG AFTER THAT, I followed through on my dream to be a Drill Sergeant, but the Army had other ideas. The personnel authorities disapproved of the request by claiming that I have not been stationed in San Antonio long enough. However, three weeks later, I was on assignment for Recruiting School instead. Initially, I was not a happy camper at the Recruiting School, but I loved Indianapolis. This was my third time there! So I put the good with the bad and went about it the right way, doing the best I could. I still hated Recruiting and felt cheated out of becoming what I thought I was born to be, a Drill Sergeant. But I learned a lot at the school, lessons that helped me succeed later in my career.

God has a plan for all of us; many times, we do not understand some of the trials, but eventually, He reveals to us why things happened the way they did. This school and assignment prepared me for bigger and better things later in my career. I am convinced that my experience as a recruiter was critical to my success as a First Sergeant of a major Army command. Once again, thank you, Lord!

You could not fail Recruiting school—it was impossible. There was no way that you were not going to pass. Maybe because very few, at least in my class, wanted to be a Recruiter. Our own census found that only three out of forty in our class applied to the school; our respective career branches RECRUITED the rest of us. It was typical for the recruiting command to tell us that we were there because we were in the Top Ten Percent in our fields; the cream of the crop as they called it. I found this odd because once we reported to our recruiting stations, they treated us like the lower ten percent. But back to recruiting school.

Things got interesting for this introvert. Our first class consisted of reaching into a bag (with our eyes closed), pulling out an item (any item), and speaking about it for three minutes. Three minutes might not seem like a long time until you must improvise and talk about something you just got your hands on. Thus, of course, you were not prepared to speak about the object. Well…I believe my "speech" lasted less than 30 seconds. This was not a pass or fail drill. It was to show us, at the end of the course, how much progress we had made and how the course prepared us for a life in Recruiting.

What was the point of this drill? Recruiting is characterized by many impromptu situations, and you never knew from day to day what was going to happen and where you were going to meet your next applicant.

When I tell you that you cannot fail the Recruiting course, I am telling you the truth—you cannot fail this course. They will not allow it. One of my classmates failed a test on purpose; at least, that is what he told us. He was trying to get out of recruiting and tried hard (like me) with no luck. So here he was, like many of us, unhappy. He took it the extra mile and failed a test on purpose. However, he was only one answer short of passing. Well, he received two hours of re-training after our regular school hours by one of the instructors (I guess it was a message from them). When it was time to re-test, the instructor showed him the question he needed to answer to pass and (since it was an open book test) where the answer was in the book.

Nevertheless, he passed the re-test for obvious reasons. He tried failing a second time and received the same help. He was not the only one who failed and got re-trained. In the end, everyone graduated from recruiting. You can say that *many even failed at failing*!

The most unbelievable thing about recruiting school was that we were trained to sell people something we could not show them. For instance, when you go to buy a car, the salesperson will show you a car and everything in and out of the vehicle, including all of the gadgets, the CD player, speakers, power this and power that. The same goes for things like boats, houses, and so forth. Although I hated being there, recruiting school was one of the best schools that I ever attended. We were trained to sell people something that we could not show, yet they would buy it and be happy about it! We were selling them an idea and a new lifestyle.

The most unbelievable thing about recruiting school was that we were trained to sell people something we could not show them.

During the final test, we met the same bag of items as at the beginning of class. This time we had to speak for about 10 minutes. When it was my turn, I spoke for 10 minutes like it was nothing; the instructor had to tell me to stop! It would not be me if I did not tell you that I basically learned how to be "full-of-it." Let me say this to those of you who know me now, you can blame it on Recruiting! LOL

I was now officially a USAREC Ranger, as the guest speaker called us during the graduation ceremony.

Welcome to recruiting—this was a nightmare that did not want to end. When I reported to the Recruiting Station in downtown Houston, Texas, they had no time to do anything for or with me. I kept asking them to give me the name of a good hotel to take my stuff there and begin looking for an apartment. This was the biggest bunch of uncooperative people I had ever met in my life, and they were going to be my coworkers. Think about that.

I spent approximately four hours at the station, waiting for somebody to guide me in any direction. This was my workplace, but I still needed to report to the battalion headquarters for processing. They would not give me an address or anything other than to sit and wait. I was then forced to spend the night with another recruiter and his family. He was a good guy and had a lovely family, but he and I had never met until that day. Thus, he was now bringing a total stranger to his house to spend the night. We left the station at about 8:00 PM to go to his house. I met his family and displaced one of his kids, and stayed in his room. His wife and children were awesome, a beautiful and welcoming family. Nevertheless, that was a day I will never forget. I began to hate recruiting, and it was getting worse by the minute.

The next day he took me to a couple of apartments. I selected one that was nice and seemed not too far from the recruiting station. At least, that is what it felt like. That is until I moved in and had to drive to work. The traffic in Houston, Texas, was worse than what I met in New York. I could practically see the building that I was driving to, yet it would take me about two hours to get there because of the traffic congestion. Many of Texas' most congested freeways are in Houston, and it just so happened that I was now living next to one of them, and it was my way to go to work. This frustrated me to the point that I had to leave my apartment at 5:30 AM to make it to the recruiting station at around 7:30 AM. A year later, when my lease expired, I moved to Houston's Mexican-American area; it only took me about 5 minutes to get to work.

Here is a picture during my days in Recruiting. Not too happy, as you can probably see. But everything was not awful.

About a year into recruiting, I had severe sinus issues and went to see a doctor, who referred me to a specialist. I was out of work for about two and a half weeks. During this time, the recruiting station commander was supposed to take me off recruiting status but failed to do so. Upon my return to work, and because my station commander forgot to adjust my mission, I had to see the Battalion Commander for missing my quota for the entire month. It was not a friendly visit with the Big Boss.

I remember that there were about 20 of us at the meeting. The Battalion Commander had us form a circle and went on "chewing out" every single one of us. When he got to me, I tried to explain that I was not supposed to be on mission because of the surgery, and he yelled at me, almost touching my nose with his, and said, "SHUT THE F**K UP. I AM TALKING NOW. YOU CAN TALK WHEN I AM DONE. YOU GOT THAT?" However, when he was done with me, he went to the next guy in line without letting me speak.

The meeting intended to make us feel like we were worthless pieces of crap for not doing our recruiting job and to motivate us to do better the next month. And such is the treatment given to the "Top-Ten-Percent," as we were called. That was a sorry "technique" to motivate; it showed poor leadership skills—to say the least. I have to give the Battalion Commander (a Lieutenant Colonel) credit for illustrating how toxic and incapable *some* "leaders" feel and act. Apparently, they lack the proper intellectual capabilities and the noble virtues to bring the best out of people. They conceal their shortcomings and incompetence behind their toxic and juvenile behavior.

It highlights the lack of leadership ability of someone the United States Army has placed the utmost trust to lead men and women into battle; although, this was Recruiting. Leading by intimidation is the equivalent of "piss–poor–leadership," or PPL, as I prefer to call it. This technique is applied in the Combat Arms fields due to their critical mission and where questioning someone's orders and authority could be the difference between life and death. However, we were Recruiters and not combat arms Soldiers. Our success relied more on our positive outlook and attitude while selling a product; in this instance, the Army.

After that "blasting," we were expected to come out of the meeting with a smile and ask the next applicant about joining this great organization. Imagine! I was now expected to ask others to join a service that I hated at the time. If I did not like this for me, why would I like it for anyone else? Think about that one, Recruiting Command.

Recruiting was not that hard, at least not for me, but not because it was an easy task. It was easy because of the training we received at the school and the additional training we received while on recruiting status. Giving you the tools to be successful was not a problem in recruiting.

The treatment that I received as a recruiter, however, was the worst throughout my military career. I liked that I could influence young men and women into serving our great nation. In addition to better yourself, serving in the Armed Forces opens uncountable doors that are unmatched in the civilian sector.

But Recruiting Command fails in placing its recruiters in the proper frame of mind and pays no attention to their overall attitude or well-being; it is all about meeting a quota. In fact, applicants could sense the dejection that recruiters reflect, at least from my experience and perspective.

Most of the recruiting's praise was shown during the Quarterly Recruiting Appreciation conference, where some of the biggest crooks were rewarded with trophies and medals. Not all of them were crooks, but the one who got most of the praise every quarter while I was there, was eventually caught on some inappropriate actions, got disciplined, and was thrown out of recruiting. We knew that something was amiss because the statistics on the quality of the applicant's education level, as reflected in the Armed Services Vocational Aptitude Battery test (ASVAB) scores in his area, were the opposite of what he was recruiting for the Army. The ASVAB is the test used to enlist in the Armed Forces. As I said, he was finally caught and paid the price with his career. When the dust settled on the investigation, we heard from those that shared the office with him that he had a copy of the ASVAB and blank High School Transcripts and Diplomas. They also found a list of college-educated "ringers" that he would pay to take the ASVAB for others to ensure his recruits' top quality. To make it crazier, they found the documents inside his desk at the recruiting station.

However, as I previously stated, the training was excellent, and when combined with your personal experiences, you could answer almost any objection an applicant could have. Although I did not want to be a recruiter, I chose to do what I was told to do.

At the time, most high schoolers were more worried about Basic Training than anything else in the Army. Their concerns were easy for me to address because I did not speak English and made it through Basic Training. Thus, I told them it should not be hard for anyone else. I paid a lot of attention while at the schoolhouse and learned how to sell the Army to others. Sometimes I resorted to challenging someone's *machismo* to get them to join the Army. It worked a few times, but only because they needed the extra push. The candidates had already decided that they were going into the military after graduating from high school.

I went to visit many candidates. And I spent many hours conducting bilingual recruiting interviews in Spanish and English in the same house. Mom and Dad did not speak English, and their son did not speak Spanish. I do not know how they communicated while living together. Many of the interviews were not with my applicants; they were with somebody else's applicants. I do not know how they did it. Still, my friend SGT Gardner (an African-American) was the recruiter in Houston's Hispanic area, and I (a Hispanic) was the recruiter in the African-American area. I spent most of my time translating for him because he did not speak Spanish, and I spoke both languages. We tried to get the station commander to swap our areas, but he declined to do it every time. He only succeeded in making our lives as recruiters difficult.

I had one applicant who practically maxed the ASVAB test. His score was so high that he qualified for every job the Army had to offer. Part of the processing to enlist in the military is conducting police checks on every part of the country the applicant had lived in. This one applicant lived approximately three hours away from the recruiting station. I told him to come with me and keep me company as I drove to his hometown to get the police report. But before driving for three hours, I decided to stop and do the local police check in Houston first. Once I parked the car, he chose to wait for me in the vehicle. I asked him if there was anything wrong, and he told me that he did not like going to courthouses and would wait for me in the car; a red flag went up.

I went inside the courthouse while he waited in the car. The police report showed that he had a long list of felonies and misdemeanors and just got out of jail a little over two weeks before taking the ASVAB. When I returned to the car, I told him to get out and not to come back. To think that I was about to drive for a total of six hours to run a police department check on someone who could not join the Army or any other branch of the military!

A big recruiting tool is a list of High School students that took the ASVAB in their respective school. The list is then provided to the recruiter assigned to that school by Recruiting Command. It contains, if my memory serves me right, the name, ASVAB Score, phone number, and the address of the individual tested. As a recruiter, you must act quickly because every military branch, the competition, gets this list. Some of the many telephone conversations could be summarized as follows:

- *If I wanted my son in the Army, I would buy him one.*
- *I disapprove of the military. Don't f***ing call my house ever again.*
- *Yes, I came to this country from South America, but I disapprove of the military. If there were to be a civil war, I would go back to South America.*
- *If you call my house again, I will go to the Recruiting Office and beat the crap out of you*
- *Only losers join the military.*

Losers? Hmm… I found it funny that many of those anti-military individuals referred to us as losers and eventually joined the military. People live in this country based on freedom; yet do not support the organization that provides it, the military.

Some parents assume that the Army should do what they failed to do while raising their children. Many times some kids end up "raising" themselves because both parents work. Here is an example, one of the prospective applicants got into legal trouble. The dad called so that we could enlist him in the Army and "fix him." Needless to say, once you get into legal trouble, you are no longer allowed in the Army. There are exceptions, but it takes a lot of paperwork and months of waiting time.

Many of my applicants were high school seniors. One of them was the President of the local National Honor Society. He was very smart and scored high on the ASVAB test. When

I called him to ask him about his interests after high school, he told me that he was planning on joining the Army. Thus, he began to collect the required documents and information, and I processed the required paperwork.

Not too long after that, he was at the Military Entrance Processing Station (MEPS), getting processed to join the Army. The MEPS was on the third floor, and my recruiting station was on the first floor. Somehow it was taking too long for him to join, so I decided to go upstairs to find out what was going on. I found him at the guidance counselor's office, looking at nothing but a long list of combat arms job classifications. When I looked at the computer, I noticed that the counselor forgot to enter the rest of his scores, and that was why he was only getting combat arms options (Infantry, Armor, Field Artillery, and others). I had worked with this branch-assignment system when I was in the human resources field—before I became an Army Recruiter. Using my expertise, I entered the rest of the ASVAB test scores on the computer.

Jackpot! All kinds of jobs in all sorts of fields came up. My applicant was happy when he saw a world of options. And I was too until the Army Career Counselor walked in and asked me what I was doing. I told him about what I thought was an oversight on his part; he got upset and told me to leave. Apparently, he put his ego first so that he could make his quota and be praised by his supervisors, totally ignoring the future of a young and promising man.

A little after that incident, I was called into the station commander's office to get lectured about what I did. He had the 1SG, Company Commander, Battalion CSM, and the LTC Battalion Commander (my "friend") on the speakerphone. I was directed not to set foot on the MEPS floor EVER. You could say that I was blasted by a Claymore mine; the Claymore mine is command-detonated and directional; simply put, it is fired by remote-control and shoots a pattern of metal balls into the kill zone, like a shotgun. Thus, I was the target of the Claymore. I was lectured—if you can call it that. I was told that the next time I had an applicant, I was to take them up on the elevator until it stopped on the MEPS floor, watch them out of the elevator, "BUT DO NOT STEP ON THE MEPS FLOOR; KEEP MY FEET INSIDE THE ELEVATOR." And, thus, I was permanently barred from setting foot on the MEPS floor until I left recruiting. All because I did the right thing for a young American.

All because I did the right thing for a young American.

The reason guidance counselors enter only a few scores on the computer is to meet their own monthly quotas. This helps them steer an applicant into critical fields that the Army needs to fill. That is done at the beginning of the Recruiting Month. At the end of the Recruiting Month, if they have not met their quota, they go into panic mode and add all the ASVAB test scores into the computer because now they are looking to reach a number. Suddenly, classifications are not important; every job in the Army is now available. Yes, you guessed it. How is it fair to give a very smart person narrowed job options just because of timing? And how is it OK to offer excellent opportunities to other applicants who are not as intelligent (test score wise) and ambitious just to meet quotas on time? I think the Army might not be placing the best person on jobs because of these practices. Now you have insights regarding what is going on inside a recruiting station and the MEPS.

Because we were not doing that great as a station, missed making our quotas almost every month, and continued to complain about the treatment we received from him, the station commander got reassigned. A new one came in within days.

His personality and caring attitude were a 180-degree turn from the one he replaced. Thank you, Lord. He tried to convince me to stay in recruiting, but it was too late for me. Plus, I did not want to be there from day one. I did not like recruiting from the beginning, and the verbal and mental abuse we received there did not make things better for me.

The previous station commander told him that I was a disappointment, but he did not feel like that and told me that he would do anything to help me, unlike his predecessor. I told him thanks, but no thanks; I was ready to go back to my field. He asked me if recruiting was too hard and if that was why

I wanted to leave. I told him that recruiting was painful but not hard. And out of the blue, I picked, without looking, five cards from my applicants' box and told him that by Friday, all of them were going to be in the Army (this was Monday). As promised, they were all in the Army Delayed Entry Program on that Friday. He tried hard to keep me, but I told him that because of his predecessor, my heart was not there anymore, not that it ever was.

Seoul, South Korea

IN THE SUMMER OF 1989, and soon after being released from jail, err, recruiting, I received military orders to South Korea. Upon reporting to the 8th Personnel Command (PERSCOM) Replacement Detachment, I was assigned to the 199th Personnel Service Company (199th PSC) in Camp Coiner, Yongsan, South Korea. The compound, as they call Army bases in Korea, was in the same location as the Replacement Detachment and in the middle of Seoul, Korea.

I was soon a member of the Niners or the 199th PSC. There was a lot of pride in being a member of the Niners sports-wise. Our biggest sports competitor was the 8th PERSCOM group, a much larger company. The competitive issue had something to do with bragging rights since we worked right next to each other. It was a friendly competition, and sometimes we combined to form one team when there were not enough players to play a sport; it did not happen too often.

As I had done everywhere else, I looked for the post softball team to try and join them. In this instance, the coach told me that he had already selected the team and would not let me practice with them. I sat on the bleachers to watch the practice since one of my coworkers was part of the team. While I was sitting there, the coach mentioned to one of his assistants that he was going to give a tryout to a Lieutenant, hoping to fill the last slot. This happened right after he would not give me an opportunity. I went on to play for our company team. I met others who did not get a chance to at least try out for the post team.

One of the guys, who went by the nickname "Doc," decided to form his own softball team and invited me to be their shortstop, which I gladly accepted. I became a member of the famous Korea Guzzlers, who up to this day is still playing softball and winning tournaments all over the Pacific, or the Asia Circuit, as it is unofficially known. The team has grown into an international organization, with teams in the U.S., Asia, and Europe. One of my teammates was Major Tom Costello; you will read about him later.

The team signed up to take part in a Pacific-wide Softball Tournament held in Seoul that year. There were teams from Japan, Okinawa, the Philippines, Samoa, and Hawaii, and every branch of the Armed Forces was represented. In our very first game, we played against the Yongsan Post Softball team. We gave them a whooping to the point that the game was not even fair. We ended up in second place in the tournament. I guess our team was a better Post Team!

You have read previously about my playing with this post team or any other team. Still, the fact is that I never really had a lot of confidence in my abilities. As others kept telling me, I did very well, but the confidence was not there, and I did not understand all of the fuss. However, playing for the Guzzlers was the most fun I ever had on a softball field.

While still in South Korea, the Morale Welfare and Recreation (MWR) organization selected a group of players, including me, to challenge the mighty team: "The King and His Court." This was a four-man team; pitcher, catcher, first baseman, and shortstop. They were America's #1 family softball entertainment, providing the perfect fundraising event for military bases, civic organizations, and entire communities. Game routines become synonymous with The King and His Court, featuring stunt pitches, feats of athletic talent, and crowd-pleasing comedy routines that stirred laughter from young and old. One could say that they were the softball equivalent of the Harlem Globetrotters.

The event was advertised throughout military bases in Korea, so the place was packed with spectators on game day. It was standing room only, and I do not recall ever playing in front of a larger audience. The pitcher was Eddie Feigner, "The King" himself, yes, the legendary Eddie. He was throwing a myriad of crazy spinning pitches. Although Eddie's pitches did not possess the speed that characterized them when he was young, as mentioned by the announcer, they were fooling everyone that came to hit before me. When it was my turn, lucky me, he decided to let his shortstop pitch.

Initially, while this "pitcher" was warming up, his fastball transformed the game from softball to 'baseball.' It seemed to me like he was throwing a 90 MPH fastball (yes, really fast). I assure you that if you blinked as a batter, you missed the pitch.

So here I am, next batter in front of the large audience. My mind racing, first pitch woosh, barely saw it. I remember thinking, *could you, "pitcher," move back and pitch from second base?* At least that would give me half a chance to catch up to the speed. Then, as fast as the first pitch, and before I realized it, the umpire called, "strike two." Wow! There I was thinking, "I am about to get embarrassed in front of all of these fans."

He then threw me a change-up. My eyes got big from the surprise of the slow pitch headed my way. I swung, and because of the enormous difference in speed between this pitch and the others, I hit an infield-fly. In other instances, hitting an infield-fly would have been an embarrassment. In this instance, I could not have been happier to hit one in front of thousands of fans.

Many clapped because it was the first time any of us made "contact" with any of their pitches. Little did they know that

I was scared to death! I faced the King later in the game and grounded out to the shortstop. I was also honored to find out that he struck out my hero, the great Roberto Clemente, during a celebrity softball game against many Major League players. I am proudly honored to have faced someone he did, even though I met and faced the King when he was much older and not as sharp as when Clemente did. After the game, a few of the kids in the audience asked us to sign their baseballs. I was surprised but obliged.

> *Little did they know that I was scared to death!*

Now let me go back to the challenges I found when I was on duty. Here is a story about doing what is right, even when you risk your career.

Our Supply Sergeant at the time was a friendly Staff Sergeant (SSG). He was always quiet but did his job to a high level. We were never short on anything. He always found a way to make things happen. Well, one Friday night, while I was on Staff Duty and making my runs as specified by the Battalion Command Sergeant Major's (CSM) instructions, I found the SSG leaving the Yongsan Post Club. It seemed like he was drunk, which I confirmed when I stopped to talk to him. He was the opposite of what I had known him to be. He was talking a lot and staggering all over. I got him to get in the Staff Duty vehicle when this other person approached me and told me that he was a Sergeant First Class (SFC), and if I did not remove that guy from the government vehicle, he would be sure to get me in trouble with my superiors. I told him to go ahead and do whatever he thought he needed to do, but I was going to take care of my fellow Soldier. He began to argue with me and tried to pull rank on me, a Staff Sergeant at the time. I told him to go ahead and do what he must do, but I was not going to follow his instructions. As I remember, he threatened to call the Military Police (MPs), but I am not sure that he did because I left with my fellow Soldier before the MPs arrived.

When I took the SSG back to our company barracks, he kept telling me, "Plablo, Plablo, I love you, man; you're my best friend, Plablo!" He kept saying it over and over. It was funny, and I was glad he was a happy drunk. That was the most I had heard him talk. Too bad he was drunk, but he was a good guy, and I was going to take care of him before the MPs picked him up and possibly cost him his career. I wrote about the incident in the Staff Duty Daily Journal because of the threat from the other guy. In the journal, I wrote that I picked up a fellow Soldier from the club and took him to the barracks, with no mention that he was incapacitated.

Once I dropped him off at the barracks, the CQ told me to go to the second floor and see what two of my Soldiers were doing. He had everything under control and was not letting them go anywhere, but he wanted me to see what was going on. The CQ led me to two male roommates. One happened to be one of my Soldiers. They were so drunk and thought they were drowning in their empty shower tub because they just had the famous "soju experience" (the South Korea national drink experience). I thanked the CQ for keeping an eye on them and keeping them out of trouble. On our next duty day, I had a long talk with both of them.

Monday morning came, and the CSM called me to report to his office as soon as possible. I thought I was in trouble because of the incident over the weekend at the Club. The CSM told me (he had yet to read the Staff Duty Daily Journal) that he received a phone call from a SFC and wanted to hear my side of the story. I told him what happened, and if it were going to get me in trouble, it was worth it, and I would do it again. I was in no way challenging his authority and said this in a very respectful tone and manner. The CSM got up from his chair and presented me with a Brigade Challenge Coin for my actions. In the military, challenge coins are used to reward team members who perform their duties exceptionally well. Receiving a coin is a big deal—it builds pride and morale. It is a great honor, and team members work hard to get them.

Monday morning came, and the CSM called me to report to his office as soon as possible.

He also told me that the Colonel agreed with what I did. He said that he went off on the SFC and was sure to tell him that I was acting under his guidance. From that day forward, all Staff Duty Officers (SDO) and Staff Duty Noncommissioned Officers (SDNCO) from our brigade were briefed to do the same thing I did and take care of each other. Only God knows how many lives and careers the new policy saved—just because this Soldier was determined to do the right thing for another Soldier.

Since my first day in Korea, I kept hearing, almost daily, about the soju experience. Let me elaborate on soju.

MOPP training in Korea or as the guy says in the famous Army parody movie Stripes, "Aaaaarmy training sir." Army fun at its best! MOPP = Mission Oriented Protective Posture

It is a clear and colorless distilled beverage of Korean origins. It is usually mixed with orange juice, and they refer to it as jungle juice. There might be other ways of drinking it. When it is mixed with juice, you could not taste the alcohol in it. I am not a drinker and have never cared to drink. But I also did not want to be in Korea and not experience something that I heard about in the States before my tour.

One day I decided to find out what the fuss was all about. I chose to spend the weekend in a hotel room in the town nearest where I was stationed to stay away from people who knew me. I chose a hotel with a bar, so I did not have to walk too far if something happened. There I was, sitting at the bar after dinner, and I ordered my first jungle juice. It was a small glass of orange juice with soju. I took my time drinking it, even though it tasted like just cold orange juice. I could not taste anything else but the orange juice, and once done with this one, I asked for a second one. While drinking the second drink, I had the urge to use the bathroom. I did not feel tipsy or anything else; however, I assure you that I could not feel the floor when I put my feet on the floor. I then stomped my feet on the floor. I could hear my feet but could not feel them

These Soldiers passed their PT test but wanted to do some additional PT. Here we are after work running in Camp Coiner, Seoul, Korea. That's me leading the run.

hit the floor. It was like I was stepping into the air but with the sound of feet touching the floor. At that time, I said to myself, *Well, I guess I do not have to go to the bathroom.* That was the beginning and the end of my jungle juice experience.

Back to sports and physical training. While I was playing softball in Korea, I trained in the gym consistently. One day, I injured my back lifting weights and had severe back spasms. As a result, I went to the troop medical clinic (TMC) for treatment. The TMC is a much smaller version of a hospital, with many located in a military base. Its primary role is to provide a full range of primary care services to the soldier population; dependents go to the Main Hospital. TMCs services include treating acute/chronic injuries and illnesses, physical exams, minor procedures, health education, immunization, and specialty referrals.

At any rate, the doctor prescribed a pain killer and a muscle relaxer. Everything was fine until, a couple of days later, I started having blackouts, and when I woke up in the morning, I had no idea where I was: Puerto Rico? Fort Stewart? Germany? San Antonio? I reiterate I had no idea where I was. Consequently, I returned to the clinic to talk to the doctor. She told me that what I had was above what she could treat, so she gave me a referral form and sent me to the Main Hospital.

When I got to the hospital and the room listed on the referral, the sign at the location read Community Mental

Hygiene Clinic. I asked the Captain at that location if I was at the right place. I told him that I hurt my back and the doctor sent me there. He said that, based on the list of his appointments, I was in the right place. I told him that I did not have mental issues and only hurt my back by lifting weights and playing softball. He gave me some forms to fill out and asked me for the medications I was taking while looking in a book. When I gave him back the forms, he told me that I had a bad reaction to the pills—what I had experienced was one of the side effects, but to go ahead and fill out a 100-question form and since I was already there, he was going to give me some relaxing techniques.

So here I was lying on this couch. He told me to close my eyes, to relax, and then started to tell me a scenario where I was in the woods, just walking around. He was sort of painting the picture. The area was relaxing. I could "picture" the wind blowing on my face when I came to a river. I could hear the water and how relaxing it was, then I sat by the riverbed for a while, taking in the scenery. After relaxing for a while, he told me that now I was in the middle of the river, enjoying the coolness of the water as it surrounded me. And that is when I jumped off the couch. He asked me what was wrong, and I told him, "I DON'T KNOW HOW TO SWIM." We both started laughing, and he told me that we were through. He also

"I DON'T KNOW HOW TO SWIM."

told me to quit taking the muscle relaxer and go back to my doctor for a different prescription.

Once my back spasms went away and things went back to normal, I was nominated to attend the Australian Rappel School in Cheju-Do, a little island south of the Korean Peninsula. Australian rappelling allows a soldier to descend with a fixed rope in a standing position while facing the ground. It is a military "assault" technique whereby a soldier is on belay, allowing them to face down during the descent and fire a weapon. Initially, it is scary as you lean on the edge of a mountain or cliff and begin to lean forward until you face down the cliff. Then it is all fun from there! Quite the opposite of the Ranger Rappel that you do butt first.

My job did not involve rappelling, but my company had some allocations, and I was fortunate to be selected to go through the school. It was a grueling week-long school, packed with all kinds of activities, such as knot-tying, climbing rock faces, beach and land an inflatable assault boat, and going across a river on a two-rope bridge. The course ends with a competition between the

Rappelling at Cheju-Do. This time was only practice and not the actual 300 ft. descent where I hurt my back.

squads. It included a 1,200-foot climb up to the top of Mount Hallasan, the highest mountain on the island, and a rappel off a 300-foot cliff. Somewhat scary, and we were wore-out when we reached the top. Imagine doing the previously mentioned events and then climbing up a mountain. However, once we reached the top of Mount Hallasan, the instructors gave us a 30-minute break to look and enjoy the view. We could see all around the island! It was scary once we got back to doing what we were to do next, rappel down, but the fear goes away as you do what you need to do. It is that Army training mentality of fear is nothing but a four-letter word.

Our instructor and I had the same rank (SSG), so he relied on me to push the team to do better. As the team members rappelled down the cliff, he left me on top of the cliff while he rappelled. He looked at me and said, "I'm going first since you know what you are doing." Halfway down the 300-foot cliff, there was a ledge where you had to take a couple of steps on a flat surface before resuming the descent.

As he went down the mountain and took the steps, he proceeded to jump out and do a spread-eagle and free fall for a few feet. I watched him and told myself, "if he can do it, so can I." Once he landed, it was my turn, and you guessed! When I got to the ledge, I jumped out and did a spread-eagle. The only problem was that the guy on belay (the "safety person" tasked with pulling your rope in case of a freefall) freaked out. Because he pulled the brakes during my trajectory, going away from the mountain at the time, he turned my body into the shape of a "V." My upper and lower body went out while my lower back (being at this time the bottom of a V) slammed into the rock mountain. Needless to say, I was in excruciating pain! I do not know how I was able to get to the bottom of the cliff; all I remember was one thing: pain.

I spent two weeks in the Army hospital back in Seoul, but the back pain and spasms would not go away. It was then that the doctors decided to MEDEVAC (Medical Evacuation) me to Tripler Army Medical Center in beautiful Honolulu, Hawaii, for further treatment.

It was soon mid-December, with snow falling in Seoul, Korea. Luckily, I was on my way to tropical Hawaii! I always wanted to go to Hawaii. This opportunity was unexpected, but hey, it really helped to mitigate my physical pain.

A funny incident happened while I waited on the gurney to get rolled into the plane. For some reason, the nurse with the *flight manifest* kept looking at me and walking away, and she did this at least three times. She finally gained courage and asked what was wrong with me. I told her that the doctor mentioned that there was a possibility that I had a herniated disc on my lower back. You could see the relief on her face, and I asked her what was wrong. She showed me the flight manifest, and next to my name, the Korean typist erroneously typed "*herniated d**k*" instead of "*herniated disc.*" We laughed as I told her that there was nothing wrong with that body part. She had it corrected, and soon enough, we were boarding the plane. Yeah, that was funny (to say the least)!

The trip was harrowing due to my condition, and I could barely feel my legs when we arrived. Because of the time difference between the two locations, I was able to see the doctor within hours of landing. He did some experimental techniques on my back, and the next morning the spasms and pain in my back disappeared. A couple of days later, I had a follow-up with the doctor and asked him where I could change my ticket to return to Korea since my orders stated a month stay and I was now healed. He told me that my orders said a month, so I needed to stay there and return when scheduled by the orders. I was not going to argue with him. He released me from his care, and I did not see him again!

Wow, here I was in Hawaii, with free room and board for an entire month. Thinking about what to do with my time there, I took the elevator to the hospital's ground floor. As I exited the elevator, a Puerto Rican friend previously stationed with me in Fort Sam

While Medevac to Hawaii, I visited the USS Arizona Memorial in Pearl Harbor. It is a sad place.

Houston entered with his daughter for a medical appointment. Of course, we were surprised to see each other and made plans to meet later. Needless to say, he came to pick me up almost every day and took me all over the island. We visited Pearl Harbor and the USS Arizona Memorial; by the way, you can still see a drop of oil leaking to the surface from the USS Arizona. Yes, since it was sunk in 1941.

A few weeks later, I was back in Seoul, Korea, in the middle of a snowstorm, and with a nice tan. My chain of command asked me whether I was on vacation or hospitalized?

Soon after this lower back accident (and its domino effect), I noticed that almost every time something terrible happened to me, the God that I was still denying put me through a good learning experience. I can say that these were situations that replaced terrible memories with good ones.

Fort Bliss, El Paso, Texas

AFTER TWO YEARS IN South Korea, it was time to return to the United States. And so, in the summer of 1991, I arrived in El Paso, Texas, for my next assignment.

Here is another softball experience. When I arrived at El Paso from Korea, I still had a few vacation days before reporting to work, so I navigated the base and asked some guys at one of the softball fields if they had a post team. They told me that the post team would be practicing in that field in about an hour. I went back to the hotel room and changed clothes to try to practice with the post team. Like in Korea, the coach told me that the team was set, but unlike the coach in Korea, he told me I could practice, and if they had enough time, he would let me hit. I shagged in the outfield during batting practice. In baseball and softball, shagging is the act of catching fly balls in the outfield during practice and not during an actual game.

I am not a tall guy, so when he let me hit, he told the rest of the team to come to the infield because they did not know me and to show me the respect that they show little guys by playing in. Initially, I was going to take some swings and spray the field, but it was time for me to become Jose Canseco and go into home run derby mode because of his comments.

What happened during his first seven pitches? I hit them all out of the park. Soon the guys moved to the outfield, as I kept belting them out. So, I yelled at them about where they were going, to come back in. Remember: I received no respect when I showed up. The series of home runs changed that!

At the end of practice, the coach tried to get me to play for them, but I told him that I did not play for guys with that poor attitude. Other things happened during practice that led me to see that I did not need to play for this guy; he was a butthole. Somehow, he found my new unit's phone number and my home phone number and kept calling. At one time, he tried to pull rank on me (he was a Captain), and I told him with all due respect that there was no rank in the softball field. He even called my company commander. When I told my company commander what happened, he told me that he would have done the same thing I did.

When I reported to work, I was assigned to the Personnel Branch. My boss was a Master Sergeant that soon had a dislike for me. Probably, because he was a lifelong recruiter that was recently reclassified to our career field of Human Resources and did not possess the knowledge that I had, although he outranked me by two ranks. I held the rank of SSG (E-6), and he a MSG (E-8). That is the only thing I could narrow it to.

My boss was a Master Sergeant that soon had a dislike for me.

On this one occasion, I helped a co-worker that was "drowning with work" and had a few deadlines. I asked him about what I could do for him. He gave me one of his most critical tasks. It took me an entire day to do what needed to be done. When I returned the action to him, he told me to give it to the MSG because I was the one that did it and should take credit for it. As soon as I turned the documents to the MSG, he told me to close the door and went off on me. I guess since he had it out for me, I handed him a reason on a silver platter. When he was done taking my head off, I walked out of his office, or he threw me out; I do not remember. I said nothing to him so as not to escalate the issue. A little later, when I thought things had calmed down, I went back to his office and told him that he just went off on me for helping a fellow Soldier. The task was not in my scope of duties. So, I told him that I guessed his guidance to me was, *"do not help others and stay in your lane."* He was not building a team but a bunch of individuals. He was speechless and embarrassed, but to me, the damage was already done. However, I still helped my fellow Soldiers but told them that they could take whatever I did for them to the asshole down the hall.

On another occasion, he also told me to do something against Army Regulations, and we sort of went back and forth on the action. I left his office, pulled out the regulation, and showed him the chapter, paragraph, and line where the regulation was precise about what needed to be done. He still wanted me to do the wrong thing, and I remember telling him that it was my job to follow his instructions, but I wanted to make sure that when the "crap hits the fan," he would stand up, be a man, and admit that he instructed me to do it the wrong way. He then told me to do it the right way and get the f*** out of his office. Such a lovely language; he led by intimidation, but I was not impressed.

No matter the circumstances, I did my job to the best of my ability, took good care of my subordinates, and participated in almost every extra-curricular activity that the company had. No matter what I did, the MSG was constantly after me and tried to catch me doing something wrong. It seemed to me like it was his sole mission. Instinctively, I sensed this and was careful with my actions and words when around him. Little did I know that he had "recruited" others to spy for him. They told me this before I left Fort Bliss. Dang, people, why the hate? This was not an affront on anyone else, and I was only doing my job to support Soldiers.

At the end of the first year, it was time for my annual Enlisted Evaluation Report (EER). As it is the norm, I listed my accomplishments on the form but feared giving it to him because of his behavior towards me. The EER is a performance appraisal. It is also used to document a Soldier's past performance and, more importantly, provide a road map for future improvement. Based on the ratings, it could improve chances for promotion or could end your career.

You can imagine my nervousness, as I was now in the hands and mercy of this MSG. After losing a lot of sleep over the following week, I was summoned to his office to review and sign my EER. I remember my stomach cramping and feeling a bad vibe due to our history. Believe it when I tell you that I was expecting the worst.

When I sat down, he spoke to me about career opportunities and the expectations of my rank. I really did not pay too much attention to what he was saying because I was expecting the worst, as I previously mentioned. Although sitting down, I was shaking on my feet. He then gave me the EER to review and agree/disagree.

I was in for a shock. At that point, it became the best EER I had ever received. The MSG then asked me if I was surprised; I replied that I was. He looked me in the eyes and told me, "I tried to f*** you, SSG Diaz, you don't know how bad I tried to f*** you, and because I could not catch you, I started to become obsessed!" He then said, "but you know what? You would not let me." He confessed that sometimes he placed me in situations that would test my mettle, but I, much to his surprise, succeeded and did not let it bother me. I mentioned that I noticed it but did not know why he was after me. He continued by saying that he was looking for reasons to catch me slipping, and no matter how small the slip, he was going to let it hurt bad. I told him, once again, that I noticed but did not know why. He did not explain the reasons for his desire to find faults in me or my work.

Regardless of his feelings or anyone else's, I took pride in doing my job the best way possible. I was always trying to be the best Soldier that I could be. That was an area where I had total control. No matter what anyone else did or said, I was getting paid to do a job, and I was determined to excel at it. Paradoxically, the MSG became one of my biggest mentors and always looked out for me. Approximately a year later, I was promoted to SFC!

In addition to hitting baseball/ softballs, I could hit golf balls very far. Here I was on the Franklin Mountains, hitting golf balls to El Paso below and in the distance. LOL

Back to Korea

AFTER ABOUT TWO YEARS in El Paso, Texas, I was on my way back to Korea. When I reported to Korea, my assignment kept changing for a couple of days until it finally settled as the Personnel Administration Center (PAC) Supervisor of an Air Ambulance MEDEVAC Battalion. I was to leave Seoul for the location the next day. I was in the elevator at the hotel, on my way to dinner for my last night in Seoul. When the doors opened, there was MAJ Costello, my old softball teammate, during my first stint in Korea. He saw me and asked if I was on vacation in Korea. I replied, "Who in the military goes to Korea on vacation?" We both laughed, and I told him that I was back on orders. He asked what unit I was getting assigned to. As soon as I told him, he said not to report but, instead, come to his old office first thing the next morning because I was going to work for him and play softball with the Guzzlers. And that is how I got to stay in Seoul for my second tour in Korea. You could say that playing softball got me a nice job in Korea.

Another funny thing happened this second time in Korea. Remember the proverb, "what goes around comes around"? One Soldier who was part of the group that used to make fun of me while I swept the AG Row by myself on Fort Stewart 16 years earlier came to Korea and ended up working for me.

Here I am wearing my new rank of Sergeant First Class.

Imagine the look on his face when he saw who his new boss was. I was now two ranks above him. When I heard from other Soldiers in the section that he was terrified and worried about what I could do to him as payback, I sat him down and told him that everything at Stewart was in the past and not to worry. Somehow that was all he did. He worried all the time. I guess my payback was that he kept looking over his shoulder, but I was never there!

This second time around, I had no problems playing for both the post softball team and the Guzzlers, although I had more fun playing for the Guzzlers. As a post softball team member, I spent a week playing softball at Kadena Air Force

Base in Okinawa. Once again, like in Germany, I felt guilty that I was playing softball instead of performing my Army duties.

Soon afterward, the effects of Army life, after the back injury in Cheju Do Rappeling School, began to show its ugly head. Depressingly, I could no longer play softball. Nevertheless, like a stubborn Soldier, I kept walking around with a loaded rucksack on my back to stay somewhat fit until I could no longer feel my right leg.

I would soon have back surgery. A month after the operation, probably naively inspired by my crazy Army Ranger friends who had the same surgery, I ran a 5k race. Although it did not hurt me in any way, this was perhaps one of the most idiotic mistakes anyone could make and a bad example for subordinates.

A few months after the 5k, I went to my primary care doctor to follow up on the surgery. When the doctor asked if I was still doing my lower back therapy and exercises, I told her that I added walking around the block, and she went off on me. She lectured me on following a plan and how much physical damage I could do to my lower back by being careless. She stated that she would not allow me to walk for another couple of months. Well, I am glad I lied and told her that I was walking and did not mention the 5k run! It is funny now; it was not back then!

It is beyond me to understand the way people think or their reason for being the way they are. While I was the NCOIC of the Personnel Actions Division, after a year working for MAJ Costello, I moved to the section across from ours to expand my knowledge in the personnel arena. MAJ Costello was a great boss and friend and had nothing to do with my desire to transfer; I only wanted to learn more.

As fate would have it, the Major I came to work for was Costello's complete opposite. This character had no compassion for the enlisted force and seemed to look down on us. It was not a problem for me as I had dealt with worse during my career. It just so happened that he had to attend a school required of his branch in the United States and soon temporarily departed. He was worried about who was going to do his job, and his boss, who happened to be a LTC, and my senior rater, told him that I should have the skills to handle his duties for the time he was gone; he would help me if needed.

During his absence, the LTC and I found over 20 policy memorandums that needed to be retyped and updated with changes in regulations provided by the Department of the Army. The memos provided administrative guidance in how personnel activities were to be conducted and processed by all military units in the Korean Peninsula units. With his assistance and my subordinates' assistance, all of the memorandums got updated and distributed throughout Korea. Upon returning from the U.S., the MAJ received a briefing from the LTC about the section's accomplishments during his absence; he did not care to get briefed by me. He promptly decided to retype all of the memorandums that were updated during his absence. When he was done, he took them to the LTC for approval. The LTC called me into his office and asked me why he was reviewing those memorandums again. I told him that I had no idea but that the MAJ did not consult me about anything and decided to retype everything that I did during his absence. He then called the MAJ into his office and told me to leave. The LTC was yelling at him and expressed a few times that he should be thankful to have someone like me that did his, the MAJ's job, and mine and kept the section running flawlessly. I felt so bad for him that I took our section out of the office area, so they did not continue to witness the chewing up of an idiot.

I continued to work for this guy, doing my job as usual, until it was my turn to leave Korea and return to the U.S. Years later, I ran across him at the gas station at Fort Monroe, VA. I came up to him to greet him, and he acted like he did not know me, so I told him, "You don't know me, so I don't know you either." It was funny then, and it is still funny now.

Here is a quick story about the crazy things we do in the Army, as we train *new* officers. Not the Army but us, the crazies in the Army. Stupid for sure. Nevertheless, one of the best things about the military is the solidarity we have. We rely on each other while away from home—missing our families and even fighting for our lives. We work hard at building trust with each other; it is our way of life. With that said, we had this new Lieutenant join our outfit. She was fresh out of Officer Candidate School. Well, to make a long story short, as we prepared to do Land Navigation training, we had her go throughout the Camp Coiner military compound from building to building, looking for a box of "grid squares" and the keys to the "impact area." If you are familiar with the military, you are aware that there is no such thing. In our case, a military map has vertical lines (top to bottom) and horizontal lines (left to right). These lines form small squares 1,000 meters on each side called grid squares. And the impact area is a designated boundary within the limits of a military heavy artillery shooting range where ammunition will detonate on impact. It is an off-limits location with no access or key to it.

We probably overdid it, as she wasted about two hours going from place to place, looking for things that did not exist. The Colonel found out and called six of us into his office. He thought it was funny but told us that we kept our game going for way too long and if there would be a next time, to keep it short. We, then, apologized to the Lieutenant and told her that it was all a game; she was a good sport about it. From then on, we made sure that she shone in everything that she did. We did everything within our power to train her well and made sure she received 100% support in all tasks.

And Back to Fort Bliss, Texas

AFTER TWO YEARS IN South Korea, I found myself back in El Paso, Texas. During this assignment, I got the opportunity to serve in a Captain position (me an SFC) and became the Adjutant to the 4th Battalion, 6th Air Defense Artillery Regiment. Our mission was to produce combat-ready warrant officers (WOBC, WOAC), platoon leaders, and battery commanders and carry out the US Army Air Defense Artillery School's security assistance training program at Fort Bliss. So here I was reviewing and updating policies that officers were tasked with implementing. This was a gratifying job, and I learned a lot about how the officer structure works. A funny anecdote from this assignment is that I became a very good friend with the Battalion Commander

This was a gratifying job, and I learned a lot about how the officer structure works.

(BC). I participated with them in many officer activities. However, we had a "snafu," not us personally but more of a funny situation.

We had a Soldier who purposely did not participate in any unit activity, by pretending that he was sick, a malingerer. It was like he was riding along in the Army, and for some reason, people were afraid to do what was required and prepare the needed paperwork to discharge him from the Army.

Trust me when I tell you that we had many conversations and training meetings to improve his overall performance. Many sergeants did their best to get him back on track without any luck. He would not participate and continued to miss typical Soldier tasks. It appeared that he was doing it on purpose because he got away with it without repercussions. I researched Army Regulations and found a reason to discharge him out of the Army and brought the information to both the BC and the XO. They agreed, and separation paperwork was submitted to the higher approving authorities. In typical Army hurry-up-and-wait fashion, we waited over a month for the approval. The XO asked me every other day because the BC could not wait to get rid of the malingerer.

When we finally got the approval, I lied to the XO to give him a hard time and told him that our request was disapproved. At the same time, one of my soldiers came into my office to ask me a question. The XO left my office while I talked to the

Soldier. When I was done taking care of my Soldier, I went to the XO's office to let him know that it was a joke and that our request was approved. He looked at me and freaked out. Simultaneously, the BC's office door was locked, and we could hear him on the phone yelling at someone. It was then that the XO alerted me that he told the BC about the *fake disapproval.* The BC was calling the personnel office and the brigade's legal office and giving them a piece of his mind.

The XO decided I should be the one to tell the BC that it was a joke between us. He wanted no part of it after hearing him yell on the phone. I replied that he should have waited for me before telling the BC. In the end, we decided to see him together, but I was the spokesperson.

I began with, "Sir, we had a snafu," and went on to explain what happened while stressing that the request was approved. I apologized and told him that it was a joke between the XO and me, but the XO had loose lips. The false information was not intended to reach his ears. But circumstances allowed that to happen. He then looked at us with a mad and happy look and said, "it's okay, Pablo, I got what I wanted, now you both get out of my office so I can call and apologize, and before I kill both of you." To this day, whenever I communicate with the XO, we both talk about "the snafu!"

During this assignment, I was privileged to be the XO for 82 days as the contingent got deployed on a special mission to

Kuwait. The BC told the brigade commander that he did not need another person and that he was confident that I could handle my job as Adjutant and the XO at the same time.

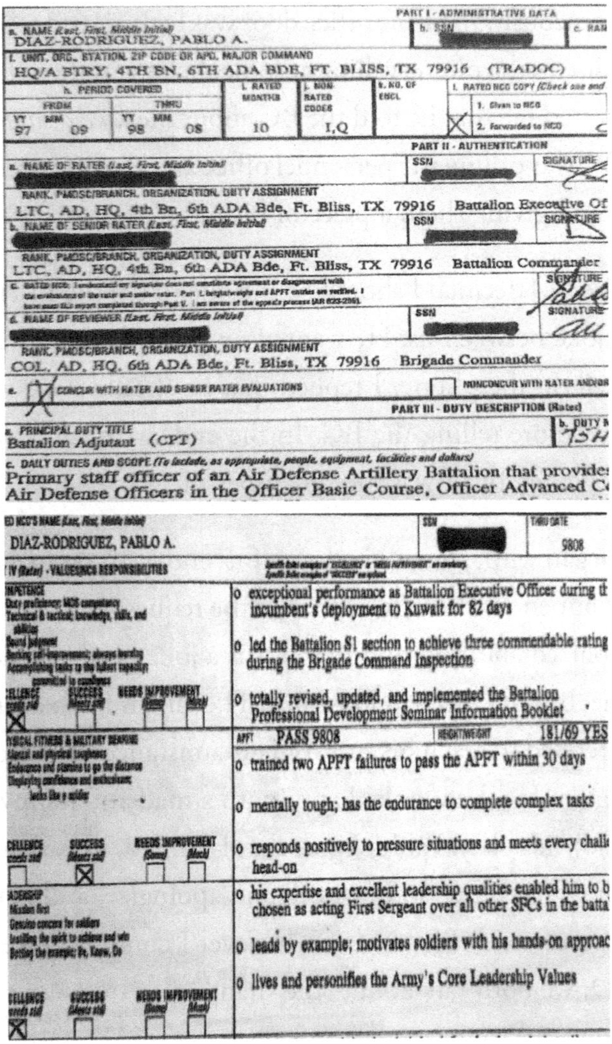

PART I - ADMINISTRATIVE DATA

a. NAME (Last, First, Middle Initial): DIAZ-RODRIGUEZ, PABLO A. | b. SSN: [redacted] | c. RAN

f. UNIT, ORG., STATION, ZIP CODE OR APO, MAJOR COMMAND: HQ/A BTRY, 4TH BN, 6TH ADA BDE, FT. BLISS, TX 79916 (TRADOC)

h. PERIOD COVERED FROM (YY MM)	THRU (YY MM)	i. RATED MONTHS	j. NON-RATED CODES	k. NO. OF ENCL	l. RATED NCO COPY (Check one and
97 09	98 08	10	I,Q		1. Given to NCO; X 2. Forwarded to NCO

PART II - AUTHENTICATION

a. NAME OF RATER (Last, First, Middle Initial): [redacted] | SSN: [redacted] | SIGNATURE

RANK, PMOSC/BRANCH, ORGANIZATION, DUTY ASSIGNMENT: LTC, AD, HQ, 4th Bn, 6th ADA Bde, Ft. Bliss, TX 79916 Battalion Executive Of

b. NAME OF SENIOR RATER (Last, First, Middle Initial): [redacted] | SSN: [redacted] | SIGNATURE

RANK, PMOSC/BRANCH, ORGANIZATION, DUTY ASSIGNMENT: LTC, AD, HQ, 4th Bn, 6th ADA Bde, Ft. Bliss, TX 79916 Battalion Commander

c. RATED NCO: I understand my signature does not constitute agreement or disagreement with the evaluations of the rater and senior rater. Part I, height/weight and APFT entries are verified. I have seen this report completed through Part V. I am aware of the appeals process (AR 623-205). | SIGNATURE

d. NAME OF REVIEWER (Last, First, Middle Initial): [redacted] | SSN: [redacted] | SIGNATURE

RANK, PMOSC/BRANCH, ORGANIZATION, DUTY ASSIGNMENT: COL, AD, HQ, 6th ADA Bde, Ft. Bliss, TX 79916 Brigade Commander

e. X CONCUR WITH RATER AND SENIOR RATER EVALUATIONS | NONCONCUR WITH RATER AND/OR

PART III - DUTY DESCRIPTION (Rated)

a. PRINCIPAL DUTY TITLE: Battalion Adjutant (CPT) | b. DUTY M 75H

c. DAILY DUTIES AND SCOPE (To include, as appropriate, people, equipment, facilities and dollars): Primary staff officer of an Air Defense Artillery Battalion that provides Air Defense Officers in the Officer Basic Course, Officer Advanced C

RATED NCO'S NAME (Last, First, Middle Initial): DIAZ-RODRIGUEZ, PABLO A. | SSN: [redacted] | THRU DATE: 9808

PART IV (Rater) - VALUES/NCO RESPONSIBILITIES

COMPETENCE — Duty proficiency; MOS competency; Technical & tactical; knowledge, skills, and abilities; Sound judgment; Seeking self-improvement; always learning; Accomplishing tasks to the fullest capacity; committed to excellence

EXCELLENCE (Exceeds std): X | SUCCESS (Meets std) | NEEDS IMPROVEMENT (Some) (Much)

- o exceptional performance as Battalion Executive Officer during the incumbent's deployment to Kuwait for 82 days
- o led the Battalion S1 section to achieve three commendable ratings during the Brigade Command Inspection
- o totally revised, updated, and implemented the Battalion Professional Development Seminar Information Booklet

PHYSICAL FITNESS & MILITARY BEARING — Mental and physical toughness; Endurance and stamina to go the distance; Displaying confidence and enthusiasm; looks like a soldier

EXCELLENCE (Exceeds std) | SUCCESS (Meets std): X | NEEDS IMPROVEMENT (Some) (Much)

APFT: PASS 9808 | HEIGHT/WEIGHT: 181/69 YES

- o trained two APFT failures to pass the APFT within 30 days
- o mentally tough; has the endurance to complete complex tasks
- o responds positively to pressure situations and meets every challenge head-on

LEADERSHIP — Mission first; Genuine concern for soldiers; Instilling the spirit to achieve and win; Setting the example; Be, Know, Do

EXCELLENCE (Exceeds std): X | SUCCESS (Meets std) | NEEDS IMPROVEMENT (Some) (Much)

- o his expertise and excellent leadership qualities enabled him to be chosen as acting First Sergeant over all other SFCs in the batta
- o leads by example; motivates soldiers with his hands-on approac
- o lives and personifies the Army's Core Leadership Values

One of my Evaluation Reports when I was Battalion Adjutant

It was a lot of work, and I worked long hours, but in the end, it was a gratifying experience. I learned the intricacies of the officers' side of the house and became a well-rounded individual because of it. Mind you, this is the Air Defense Artillery side of the Army, and I am a personnel guy, never exposed to this sort of environment.

As a reward, I was nominated by the BC, with the Brigade Commander's approval, and received the prestigious Honorable Order of Saint Barbara. The Order of Saint Barbara is a military honor society for both the US Army and the US Marine Corps Artillery, including Field Artillery and Air Defense Artillery. The award is named for Saint Barbara, the patron saint of artillerymen. It is awarded to those who have demonstrated the highest standards of integrity and moral character, displayed an outstanding degree of professional competence, served the Artillery with selflessness, and contributed to promoting the Artillery branch.

I was in such awe and could not believe that I, the enlisted admin guy, was now treated as one of them in the Air Defense Artillery field. I went from a non-English speaking Private that was not going to last four years in the Army to receive this prestigious award due to my performance as a Captain/Major.

I soon got the opportunity to serve as a chaperone to the International Military Students Office (IMSO), who were part of our battalion, as they went on a tour to Washington, DC.

By this time, I had learned a few Arabic words from the civilian in charge, who became a good friend. One of those words is *yallah*, which translates to "let's go" or "come on." During the tour, one of the students, an officer in a foreign Army, no matter what we did or where we went, was always lagging and making everyone wait and, in turn, late. This one time, we were pressed for time, and who was dragging again? You guessed it. So, I said to him, *yallah, yallah*. He then decided to get ignorant with me and started barking at me in his native language. I could not help it; I started barking at him in Spanish. If he could do it, I could do it just as good or even better. It worked! My Spanish stopped him in his tracks and shocked him to the point that he went to the civilian in charge to complain that I disrespected him. The civilian in charge came to me with the student and asked me what happened. I told him. He then turned around to the student. Much to my surprise, the civilian in charge started yelling at the student in Arabic and told him to apologize to me in English. Funny, funny, funny, and he did not make us late anymore!

> *I went from a non-English speaking Private that was not going to last four years in the Army to receive this prestigious award due to my performance as a Captain/Major.*

Fort Monroe, Virginia

IN 1999, AFTER A year and a half in Fort Bliss, Texas, I was nominated to be a member of the prestigious Department of the Army Level, Personnel Management Assistance Systems (PERMAS) Team located in beautiful Fort Monroe, Virginia. Fort Monroe, Virginia, was a very historic military base and the headquarters to the United States Training and Doctrine Command (TRADOC) under a Four-Star General's leadership. However, the day-to-day operations were led by a Company Commander, a Captain, and a First Sergeant.

At the time, the mission of the PERMAS Team was to inspect TRADOC installations for compliance with Army personnel management regulations, principles, procedures, and concepts. The team, led by CW4 Valerie Lubin (one of the brightest personnel officers I have ever met), also conducted senior management level briefings and reported evaluation findings to Installation Commanders and Senior Department of the Army Staff members. It was an extremely critical task and essential to the quality of life and support given to all Soldiers, regardless of rank, position, and location Army-wide.

A little over a year later, I was selected and promoted to the rank of Master Sergeant. With the promotion came the opportunity of being a First Sergeant, a personal goal of mine from my days in Germany and under the leadership of 1SG Booker; he was very inspiring. Here is the difference—a MSG is an E-8 in a staff or administrative position, but a First Sergeant or 1SG is an E-8 in a leadership position (and a tough one).

> *With the promotion came the opportunity of being a First Sergeant*

However, this 1SG position was like no other as this was a Major Command. It was even more challenging. Your typical 1SG handles approximately between 120—140 Soldiers. In this position, we had over 850 Soldiers of all ranks, from Privates to Generals, in addition to another 400 that belonged to the Supreme Allied Commander, Atlantic (SACLANT) Command, another Four-Star General command, that we could not take credit for but had to support.

Imagine overseeing a group of Soldiers where over 75% of them are above your rank. Yet, we had to "tell them" what they needed to do. It was a major challenge, and this is where

Change of Responsibility Ceremony.
My first day as the First Sergeant.
To my right is CSM Browning, who was instrumental in my walk with Jesus and my duties as First Sergeant. He was a tremendous mentor.

my training as a Recruiter and my previous position as a Captain/Major came in handy. It was not what we told them, but HOW we told them that made a difference, and, in turn, we received their cooperation.

We had close interaction with all Soldiers in addition to their spouses and children. Many times the Soldiers' spouses called us for assistance before speaking with their spouses since their issues were not related to the military, but more about the living conditions in the base. Let me use a comparison to illustrate both jobs (the Company Commander and the 1SG). The Company Commander, my direct boss, was the City Mayor. I, the 1SG, was the City Manager. Both jobs were extremely rewarding yet stressful.

There are many good stories worth telling about this place (so many that I will save them for another book), but one that always comes to mind is the Red Bull story. When I was the First Sergeant of the Headquarters Company, we had to do many administrative things to get the unit to follow Army Doctrine. Because it was top-heavy, as we call units with many high-ranking officers, many of the routine tasks expected of Army units were not enforced by the previous command; that was an uphill battle that the commander and I eventually won. But not without a lot of pain.

One of the officers in my company was the Aide de Camp to the TRADOC's Commanding General, General John N. Abrams. The General was a hard but fair officer and always took good care of us.

Back to the Aide de Camp. Because of the General's hectic pace and a super busy schedule, the Aide, Major Larry

With Major Larry Perino at a function hosted by his boss, the TRADOC Commander, General John N. Abrams. Major Perino later retired from the Army with the rank of Colonel.

Perino, did not have the freedom or flexibility to take the PT test during scheduled hours because of his job demands. I understood this. Well, I worked with his schedule and settled into meeting him at 9:00 PM outside the base's gymnasium, where I administered the PT test to Major Perino. He was into fitness and did well on the PT Test. Because I worked with him, he felt the need to buy me dinner at a later day. It was not necessary, but he insisted.

Once that day arrived, we walked from the base to an off-post restaurant. While there, Major Perino ordered drinks for us. Red Bull with vodka. He told me it was good and that he had it many times before. We ordered dinner and chit-chatted for a while. By then, we were eating and were done with the first Red Bull and vodka, so he ordered a second round. As we were drinking the second round, and for a reason that only he could know, he told me that he read about two guys in Germany who died from drinking Red Bull and vodka. I looked at him, puzzled, and asked, "Are you trying to kill me?" He started laughing and told me that the guys drank five in less than 20 minutes, and it gave them a heart attack. "Whew!" I said. That was one of my experiences with the Aide to the legend—General John N. Abrams. Major Larry Perino is a hero on his own. You should "google" his name and the book "Black Hawk Down."

Picture this…the Common Tasks Test, or CTT as it was called in my day, was a series of, as the name says, common tasks that every Soldier had to be tested once a year. Fort Monroe was one of those locations where CTT was not "observed." That is until I became the First Sergeant. I was not trying to be a hard ass, but no matter where our duty location was, we were still Soldiers, and soon enough, we were due to leave and go to another Army base. Keeping proficient at being a Soldier was a must. The Commander's responsibility is to ensure that the unit training plans prepared the unit for war by ensuring that Soldiers remain proficient in their MOS and skill level tasks. It was my job to implement the plan and ensure that everyone complied.

Well, this one time, as we got everyone ready for the CTT, the boss lady from the Army Community Service (ACS) called to tell me that they were expecting more people at their quarterly meeting than what showed up. Because of this, she had an overabundance of goodies, such as muffins, donuts, pies, cinnamon rolls, orange juice, soda, water, and many other things. She told me that if I wanted the goodies to have someone pick them up; otherwise, they would get wasted. I told my Operations Sergeant to go with another Soldier to the ACS building to pick up some important supplies. I wish I had been there to see his face when he called me to inquire about the plan and what we were going to do with all the goodies. I told him my plan, and he responded with, "Top, have you lost your mind? CSM Browning, the Post CSM, is going to kill us."

I cannot imagine any other unit in the United States Army where Soldiers taking their CTT were rewarded with such goodies as soon as they completed their test. The Post CSM showed up to check on the training, and as I was giving him a briefing about our CTT stations, he glanced to my left and saw the tables set up with all the food. He looked at me and said, "I'm not even going to ask'" and began to laugh hysterically. Then he said, "I can't even be mad at you; you are always looking at ways to make the Army more tolerable. These kids (talking about the Soldiers) will never forget the day they took a CTT, and at the end of the test, there were donuts, muffins, cinnamon rolls, and everything else waiting for them. Something that they will most likely not see again during the rest of their military careers."

This is, by the way, the same CSM that led me to Jesus. CSM Anthony Browning, a fantastic mentor that always, no matter the situation, had a positive outlook in life and was always smiling. I wanted and needed that. At this time of my military career, I was highly successful but miserable, and I knew that something was missing. But what was it? I frequently asked myself. However, I saw CSM Browning's outlook and wanted that. There was something about him. Perhaps he reflected tremendous spiritual strength.

One day, I pulled him to the side and asked him about his secret. I was surprised to hear about it; it was Jesus. I wanted to have that feeling and outlook in life.

> *One day, I pulled him to the side and asked him about his secret. I was surprised to hear about it; it was Jesus.*

Unlike Bible Thumpers—the ones that want to shove the Bible and its teachings down your throat—he softly guided me along what I later identified as the right path. However, it was a gradual process. He did not push me nor attempted to make me understand "the new life" overnight.

He gave me a book, sort of like the Daily Bread, with short stories to read every day with the Bible on the back of it for quick reference. It took some time, but I am now on the right path.

I am not a religious nut, but rather a Jesus follower trying to live the best Christian life possible. I have found hope, joy, and *real* life. My mouth gets in the way often, but I catch myself and get back to where I need to be. I try to live by the saying, "God wants spiritual fruits, not religious nuts." Now I am happier than ever before.

Columbus, Georgia and Phenix City, Alabama

IN 2004, AND AFTER 24 years in the Army, I decided that it was time to do something else and figured that by now, I could defend myself speaking English in the civilian world. Hence, I retired from the Army and moved to Fort Benning, Georgia, to help my brother's family, as he was deployed to Iraq at the time.

I found my job-hunting experience very disheartening and discriminating. I could not get a job because I was overqualified; yes, they told me this. I went to an interview with the City of Columbus, Georgia, and as I was waiting for the interview, the guy I was waiting for was laughing while looking at some papers. When he finally called me in, he told me why he was laughing. He had heard about overqualified applicants but had never seen one. He then said to me that he would not hire me because not only was I better qualified than him for the job, but I was also better qualified than his boss, who was in charge of the entire section. Unbelievable. Either you do not have enough experience, or you have too much. So off I went looking for other jobs.

My following interview was also with the City of Columbus, but this time with the Parks and Recreations Department. They had a supervisor position where the person was tasked with ensuring that the softball/football fields were ready for any event. I applied because of my baseball/softball background and figured this is something that I would enjoy doing until something better came along. I interviewed and got the job. On my first day on the job, the guy in charge decided to drop me from that position to a laborer position and with over $5.00 cut in pay. The first task was to get on a truck with a crew and pick up trash all over the City of Columbus. Nothing wrong with being a laborer; that is where I started before joining the Army. When I asked him what was going on and what happened to the position that I interviewed for, he told me, "Here, everybody and anybody starts at the bottom." I looked at him and told him to look at my resume and that after 24 years in the Army, I did not consider myself to be just anybody. He could keep the job and did not need to pay me for the two hours that they had us sitting around doing nothing until they were good and ready to put us to work.

It did not stop there. I applied for two positions with a contractor at Fort Benning, GA. One was a General Clerk and the other a Shipping and Receiving Clerk, although I knew that I was better qualified to do other jobs. When I followed up on my application, I was told that I was not selected for either position. As I applied with other companies in Columbus, GA, I soon received a call from a company's Human Resources Director out of their corporate office in Houston, Texas. The position was in my field of Human Resources and at Fort Benning, Georgia. The lady asked me a few times if I wanted the job because my qualifications were above the position. I told her that I was ready to work after recently retiring from the Army. I got the job.

When I reported to work, little did I know that it was at the same place that I applied to earlier and did not get the Clerk and Shipping and Receiving positions. Imagine that! Curiosity led me into looking for the file with older job applications to find mine. Imagine my surprise when I saw "NOT QUALIFIED" on both of them. Hmm, I said to myself. I did not qualify to be a clerk, but I am now THE NEW HUMAN RESOURCES MANAGER. Think about that for a minute. I was soon to find that it was their norm not to hire people who were qualified, and instead, they were hiring people with zero qualifications and their buddies. Why? Everyone talks

"NOT QUALIFIED"

about "fairness" and "taking care of veterans" . . . they made too many sacrifices for us (more about that later). However, very few walk the walk. The fact of the matter is that there is a lot of hypocrisy in this world.

The new HR Sheriff in town was soon to fix some of these problems. I can write a book about this position, but it will probably be well over a thousand pages. It would make for a good television show, you know, those with a lot of drama. And yes, looking back, it was not funny then, but it is now!

While working there, I met the most gorgeous woman I have ever seen, who was soon to be my wife, and moved across the border to Alabama, approximately 15 minutes from Fort Benning.

Not long after I moved to Alabama, I met my wife's family. I will never forget the time when we went to visit her grandparents on a Sunday afternoon. Royce (Gran) and Verner (Anmama) Millirons are some of the nicest people I met. Very hospitable and caring. During this visit, I spent some time speaking with Gran, just the two of us, while my wife talked to Anmama in another room. We must have talked for a little over two hours in his living room.

After we left their house and on the way home, my wife asked me, *"What did you and Gran talk about?"* to which I replied, *"I have no idea, I did not understand a word that he said!"* My wife responded with, *"You mean to tell me that you both spoke for over two hours, and you did not understand anything? That is funny!"* My response was, *"yes, we spoke for that long, and I had no idea what he was talking about."* But wait, it gets funnier. After we left, Anmama and Gran had a similar conversation. She asked Gran about what we talked about, and he said, *"I don't know. I have no idea what that young'un was talking about!"* Think about that for a minute. We spoke for that long and kept talking as if we understood what the other was saying and kept talking in an effort not to disrespect each other.

Later, as we got familiar with each other and understood our different accents, we had fascinating conversations. I also learned that every time Gran told me, "*Is that right?"* he had no idea what I was saying because I was talking too fast. I know that sometimes I tend to speak very fast to the point that I don't even understand what I said! You can say that I talk like I eat, quick, pronto, and in a hurry. Yes, it is funny, and you can laugh.

As Gran and I spent more time together, we both concluded that although he was born and raised in the U.S. (and of course, I was born and raised far away, in Puerto Rico), we had similar upbringings. How is that possible? Well, for instance, we both like to eat fried chicken heads, and we also like to suck on chicken bones after eating chicken soup. That is only a couple of many similarities. He had a great sense of humor with a contagious laugh that made others around him laugh. I guess we both are country boys -Gran from the United States deep south, and the guy writing this book from the deeper south: the island of Puerto Rico—ha!

Gran is no longer with us, but his spirit lives on! He made a difference in my life. Furthermore, he and Anmama gave me the love of my life. I am forever grateful!

Here's a picture of Verner and Royce Millirons. Anmama and Gran.

Another story. I went to help my wife's stepdad Bruce, whom I affectionately call "Poppy Man," at their family farm over the weekend. After listening to his instructions and still thinking like a Soldier, I quickly gave him another possibility (faster) to accomplish our task; you know what they say: once a Soldier, Always a Soldier. After I was done explaining, he looked at me and said, "You can do it that way fyonto." I looked at him curiously and asked him to repeat what he stated. He then repeated that word 'fyonto.' I had no idea what he was saying until he broke it down and said, "If you want to." After that, I was scratching my head and thinking, *Dang, I have to learn how to speak English all over?*

While going to college and taking Principles of Microeconomics, I learned that my opinion was wrong. For real! My first test consisted of 20 questions, 18 multiple choices, and two essay questions. I did not doubt that I got the first 18 correct, as the graded test showed later. The two essay questions were a different story. They both started with "in your opinion." So, I gave "my opinion." To my surprise, when I received the graded test, it was a big fat "**F**." Imagine my surprise when I saw that. When I emailed the professor to inquire about the grading, she told me that my opinion was wrong. I argued that it was "my" answer and how could it be wrong when she asked for "my opinion"? I explained that we all see things differently from our perspective. She did not want to hear it and did not give me any points/credit for my answers. Those two answers carried a lot of weight and negated the other 18 correct ones. So much for "my" opinion and "my" answers. I suppose that Professors should be more accurate when they formulate their exams or communicate. After all, they should be the epitome of accuracy and intentional communication. Well, she did not even have the patience to listen to my feedback. Can you believe that?

Things happen to me that you cannot make up; as mentioned, some funny and some not so funny. Whenever my preacher and mentor, Brother Ricky Ward, went on vacation or was absent for any other reason, he would let me preach at church. Most of the time, I practiced my sermon in my car on the way to work and on the way back home from work. However, this one time, the Lord told me to practice in front of my wife. I did not like doing this because I did not want her to know my message before service, and she also wanted to wait to listen to it at church. But for some reason, God wanted me to practice in front of her, and I am glad I listened. As I was practicing, I made a statement to the effect of "Christians need to focus..." But that was not what was coming out of my mouth; what was coming out was "Christians need to f**k us..." No matter how many times I tried the right word would not come out, so I had to adjust my message and change it to "Christians need to concentrate..." Needless to say, a few weeks after that incident, I was preaching once again, the word 'focus' came out of my mouth. I scared myself! I quickly looked at my wife, who gave me a thumbs-up...praise the Lord!

My wife is an artist; she has drawing skills, among many others. For the cover of this book, I wanted to involve her, so I told her to make a couple of front pages (book covers), so I could select one for the book. When I told her the book's title, she looked at me inquisitively and asked for the title again. It seemed that I was saying "**Lust** in the Translation" instead of "**Lost** in the Translation" ...the stories continue.

What Does It All Mean

THERE IS A SCENE in a very famous movie where the main character takes an IQ test to determine where he should be placed academically. After the test, the school principal tells the mother where her son falls while pointing to the center of a graph labeled "State Acceptance/Normal." He explains that normal is above the red line on the chart. He then proceeds to point to a section below the line where the main character falls.

I can picture the same conversation between my parents and teachers when I was very young. The teachers showed a graph to my parents while indicating that I was all over the place, over the line, under the line, to the left, to the right, I mean ALL OVER the place. And then saying, "Mr. and Mrs. Diaz, your son is special."

"Mr. and Mrs. Diaz, your son is special."

I have always been an unorthodox person with this insatiable quest for knowledge, always asking many questions about everything. Thinking like the norm has never been a trait. It is my normal to always look for different ways of doing things. This different outlook in life may have helped me cope with the situations I encountered.

I sometimes sit back and think about why things happened the way they did and how I could have made changes to make my life better and help others. Many of the misunderstandings and unfortunate situations that I encountered taught me to see the other person's side.

Things were not easy, as you had read above, and I learned a lot during the trials. I succeeded many times by staying ahead of the game, anticipating what was next, and figuring out how to do things independently.

God was with me all of this time, even though I was ignoring Him. Due to my circumstances and the discrimination that I went through, I promised myself that I would not treat people the same way as I was treated when I got to a position where I could influence others. During my military career and subsequent civilian career, I intended to treat others as adults, and I believe I succeeded.

Another issue beyond my understanding is how it is acceptable to use certain words in conversations with individuals of our own race or culture. Still, when members of other races or cultures use them, we get offended? If it offends you when others use them, quit using them yourself.

One example about words and catchphrases not to use is "God D******." This phrase is used extensively not only in the Army but also in civilian life. Why is that not offensive? Trust me when I tell you that I used it quite frequently until I learned its true meaning. And, then, I could not find a hole deep enough to bury myself. I am also thankful that God's love for us is unconditional; else, we would be in a lot of trouble. So, Lord, I ask, again and again, forgive me. Yes, I know your mercy, just like your power, has no limits.

Another word is the infamous "N-word." This is a word that I heard my roommates calling each other at Fort Jackson, SC, so much that I thought it meant buddy. I listened to that word throughout my entire military career. We all use those types of words within our races, thinking that it is funny because it is "us talking" with each other and *keeping it real.*

Does it make sense? If you want respect, behave and speak to inspire respect. By the way, notice I wrote, "respect"—not fear. Big difference. Respect is long-lasting. Fear is temporary.

I find it funny that native English-speaking people tend to speak loudly and slowly to others who do not speak English as their first language. It reminds me of the funny hillbilly letter, *"I am writing this slow because I know you can't read fast."* Or make fun of our pronunciation as we learn to speak the language. Because of this, sometimes, my goal for the day was to avoid everyone with the goal of not speaking so they would not make fun of me and my accent. I guess if you do not speak good English, you are deaf or display slow cognitive abilities. I also find it interesting that so many native English-speaking individuals expect *the whole world* to speak proper English.

There are a lot of things that I still do not understand. There was only one color in the Army, and diversity was the norm during my time, and most likely still today. We have a saying in the Army; we are not black, white, brown, or yellow. Our color is green, and this is what brings us together to form a team. One could say that the military is strong because of its diversity—many races fighting together for one cause. We rely on each other regardless of our race and background. Diversity brings many ideas to the table that make what is already a good situation, or plans, better. The military would probably fail if it were not for its members' diversity and points of view. Although there are still issues in the military, because nothing is perfect, there is acceptance and real diversity.

We have a saying in the Army; we are not black, white, brown, or yellow.

We were all created by God with our unique purpose in life. We assume that this is all there is. Ponder the question: how can we compare 80, 90, or even 100 years of this "life" compared to all eternity? Can your imagination take you there? Make an investment where it counts—in the afterlife. Believe me. The trials are nothing compared to the eternal reward.

The trials are part of the teaching to prepare us for the next stage, as I learned later in life. The experiences and tests allowed me to learn valuable lessons that assisted me when counseling others. I hope to have saved someone's career; I know that the lessons learned saved mine.

It took an experience at Fort Knox, Kentucky, to find that all Puerto Ricans carry knives. During the in-processing, the authorities directed a Spanish-speaking Sergeant to ask me for mine. I never had one, at least not for the purpose they implied because of the stereotype. I also did not know that I was not supposed to like Mexican-Americans. Imagine that—yet every time I needed a translator or interpreter, a Mexican-American came to the rescue. By the way, in Puerto Rico, the Mexican presence is quite prevalent in the Arts and other areas; many have moved there to spend the rest of their years in the "Isla del Encanto." I wonder. . . why?

That being said, let me go over some of the great experiences I had with Mexican-Americans in the Army. Go-Go, the guy who used to sit next to me during my time at Fort Stewart, was Mexican-American, and we got along great. He taught me a great deal about the Army. The guy who invited me to play in the Negro Baseball League was also a

Mexican-American; he was very helpful and translated many dugout conversations during games. There were so many more throughout my career that were always there for me. I have never watched the film *West Side Story,* but I guess I need to watch it to learn to behave like Puerto Rican characters in the movie; apparently, that is how I am supposed to act. It is the stereotypes we see in movies that we believe. Why? Where is the critical thinking?

I cannot stress enough about the one area totally ignored by the military: the end of your service time/career. Simply put, I believe that much more can be done to reintegrate a Soldier into society. In Basic Training, they transformed us into fighting machines. That is necessary; however, there is no reverse transformation (or retransformation), enabling you to adapt to the civilian world—a world that does not understand why we are so direct, strict, organized, disciplined, and structured. There is no off-and-on switch, and no one in the military chain tries to transition and prepare you for civilian life. There is no "re-reset" button. The military system does an excellent job of indoctrinating you into the military, but a poor one at transitioning you into civilian life.

I was fortunate to have supervisors like Christine Barker, Marty Kunz, and Ed Taylor that allowed me to be a Human Resources Manager and guided me along the conversion path from military to civilian. I am also very thankful to both Pat Kenny and Bart Wilson, who allowed this unorthodox person to try different things and ideas to make things happen at the worksite.

Another critical issue today is the difficulty many veterans have adjusting to civilian life because of the lack of services to assist in the transition to civilian life. Some organizations profess to do just that. However, many are only after the money they can get from the military, and veterans' rehabilitation is just their front. I have spoken with many veterans, and they had the same experience as I did. Often, we are scheduled to meetings to help in our transition to civilian life that gets canceled at the last minute—no assistance with transitioning to a different world.

But there are also good organizations that are there 24 hours a day to assist, some free, like the Hope Project in Fountain, Florida—led by former Army Chaplain LTC(R) David Trogdon. They offer many services, all free of charge, to members of any military branch and first responders. I found some comfort there after many years of struggling.

I always felt that I was held to a higher standard throughout my life by those that were part of my professional life; somehow, more was expected of me. I was never their first choice for whatever reason, but once I began to work, all of the heavy workloads were laid on my shoulders.

Growing up, my brothers and sisters always expected me to be the strong and serious one. Many times my dad treated me differently than my siblings. He told me that I reminded him of what he should have done but did not. I learned this from my conversations with him towards the end of his life.

My parents did not raise us to discriminate or dislike others. And as kids in the neighborhood, we took everybody in. The color of your skin was not an issue. We were raised to respect people regardless of their background, race, and income level. Treat everybody with dignity and respect, my parents used to say (of course they said it in Spanish!), and God will watch over you.

I learned from my mother the importance of making do with what I had. As alluded to, baseball was my everything growing up, and I wanted nothing to do but to play baseball after school. My love for the game was so strong that I would ride my bicycle for miles to the next town, looking for pick-up games. I used to get highly upset when it rained, and I could not play baseball. In school, I would answer my homework questions in my "libreta" (notebook) so that as soon as I got home, my mother could check it, and I could go play. To save time, I used to do my homework during lunch breaks when I had time left. Sometimes I would be walking home, answering some of the questions I could not answer in school. As soon as I got home from school, I made some coffee, as it was my mom's and my personal tradition; I brewed coffee while mom checked my homework.

One of those times, the filter that we used to make the coffee had a hole, and I could not use it. I asked mom if she had another one. Well…much to my surprise, she told me to get a clean sock. Yes! You should have seen the look I gave her. She told me to get "*una limpia, no una sucia, y esta seguro que no es de tu hermano mayor,*" translated to "*get a clean one, not a dirty one, and make sure it is not one of your older brother's.*" His feet stank so bad that even when cleaned, his socks still reeked. She then said, "*muchas veces tenemos que improvisar con lo que tenemos,*" which translates to "sometimes we have to make-do-with-what-we have."

Above all, my parents taught my siblings and me loyalty and respect. These values helped me to throughout my life in the US Army. However, loyalty and respect are lost with the new generation. They are more entitled than anything I have ever seen. But we do not need to blame them. Their behavior is something that we did ourselves; we created this monster. They are the result of what we taught them and allowed them to get away with. We need to fix ourselves first.

Many things happened to me, but I also had my share of things that I am not proud of. I offer my most sincere apology to those I mistreated or insulted in any way, shape, or form.

I assure you that, after a reckless life, I finally learned the hard way. I ultimately chose Christ. He healed me not only spiritually but also emotionally—and better than many psychologists. He continues to transform me. However, I can say that I have found purpose, hope, and genuine happiness in Him.

I ultimately chose Christ.

I would be remiss if I did not mention my friend Jose Marrero's FLAG definition; it stands for **F**amily, **L**ife, **A**merica, and (above all) **G**od. All members of the armed forces, be it Army, Navy, Air Force, Marines, and Coast Guard, go out every day to defend just that. Many countries worldwide are jealous of our freedoms and seek to destroy our way of life. We have enough enemies. We need our military.

A word for those in despair—or close to that point. There is no reason whatsoever to give up on hope. After every storm, there is sunshine; after every night, there is day.

If I, miserable and unfortunate, was at the end of my rope and, much later found joy, peace, and happiness, then, anyone can have a new day full of hope—sooner or later.

"Never, ever give up on hope. Never doubt, never tire, and never become discouraged. Be not afraid"—Pope Saint John Paul II

I assure you, Christ's mercy is by far more extensive and more robust than our sins and wicked behavior. He always receives with open arms anyone who feels sincere repentance.

Here is another excellent point to ponder:

"The world will not be destroyed by those who do evil, but by those who watch them without doing anything."—Albert Einstein

Until the Next Time

THERE ARE MANY MORE stories not included in this book. Some of them are still not funny, even after all these years. Some of them scarred me for life, to the point that I needed to get professional help. I am better now, and writing this book was therapeutic, as I could look back to those days and laugh. I hope you enjoyed the stories and laughed too! And I leave you with the following Bible verse, which I find very helpful during those times that I am facing situations that have the potential of bringing out the worse in me.

"Do not be overcome by evil but overcome evil with good."
Romans 12:21

Made in the USA
Monee, IL
04 November 2026

67379289R00105